THE IMPACT OF INNOVATION:

ENTERTAINMENT

I0471293

Tom Stuczynski

The Impact of Innovation: Entertainment

THE IMPACT OF INNOVATION: ENTERTAINMENT, FIRST EDITION
Copyright © 2014 by Tom Stuczynski

Edited by Alexis Santos
Cover art designed and created by Vanessa Maynard

For more information, direct all inquiries to:
contact@impactofinnovation.com

Table of Contents

Preface

The idea of writing a book first came to me in late 2011. As an avid technology enthusiast, I pay close attention to the latest news and trends in the industry; especially those areas that have potential to make a major impact on the future. However, I realized that coverage of these exciting innovations was minimal outside of technology-oriented news websites.

After further investigation, I discovered most people just aren't aware of the awe-inspiring capabilities of technologies that exist today. With this in mind, I set out to write one book to highlight these innovations and describe how each technology could evolve in the years ahead. The hope was to generate awareness about where we are technologically and where we may be headed.

It wasn't long before I realized that there were far more topics to discuss than what would ever comfortably fit in one book. I decided to split the manuscript into a series, with each book detailing the technological progress in a single area of society. This book, which is focused solely on entertainment, is the first of many in this series.

The greatest challenge faced while writing this book was keeping the work as up-to-date as possible. Since the topic at hand is about cutting-edge technology, it wasn't uncommon to finish writing a section only to discover shortly thereafter that a major event had occurred in that subject area. Even though I've given my best effort to keep the content as fresh as possible, I've conceded that some of the information in this text will undoubtedly be outdated by the time it's in your hands. This was the driving force behind the companion website (*www.impactofinnovation.com*), which I intend to keep updated with the latest developments in areas discussed throughout this text.

The primary motivation of this book is to shine a light on current technological trends in the entertainment industry. In addition, it's my sincere hope that this text will spark a few conversations among friends about how these technologies should develop. After all, understanding the potential impact of a new technology is only the first step in minimizing undesirable side effects. We must also work towards reducing those unwanted impacts. I have provided numerous discussion questions at the end of chapters to promote such conversations. There is also a discussion board on the companion website to serve as a platform for conversation. I encourage all readers to visit the website to learn about the latest updates, submit content to share with others, and discuss ideas about what the future may hold.

Researching and writing about the latest technological trends has been an absolute pleasure. In the end, if readers enjoy the book anywhere near as much as I've enjoyed writing it, then I would consider this endeavor a great success.

"There's a way to do it better - find it."

- Thomas A. Edison

Chapter 1: An Introduction to Entertainment Technologies

We live in a time when video games can be played with nothing more than your mind. Merely thinking about an action can cause it to occur on screen. While it's true that this type of technology is in its infancy and is not yet widely available, the concept is nonetheless a reality.[1]

Let that sink in for a moment. We've reached the point, technologically, where brain activity can be used to communicate with electronic devices. Such an extraordinary concept was limited to the pages of science fiction novels just a few years ago, but is now considered science fact. What's more, this concept is on the verge of forever changing how video games are played.

Although this progression, by itself, is quite incredible, the truly remarkable aspect of this innovation is that this type of awe-inspiring technological advancement is not a rarity.

Scientific breakthroughs are occurring nearly every day that are making some of the most exciting and ambitious technological concepts possible.

---

[1] Mind-Controlled Videogames Become Reality." *The Wall Street Journal.* Dow Jones & Company, n.d. Web. 19 Apr. 2014. <http://online.wsj.com/news/articles/SB100014240527023047076045774262 51091339254>.

Today, musicians can create music by simply moving their arms and shaping sound waves with their hands. Sports equipment is becoming smart enough to communicate information useful in reducing injuries. An entire library's-worth of books can be accessed at a moment's notice from nearly anywhere in the world using only a mobile device carried in a pocket. All of these innovations are extraordinary, but they are only the beginning.

Readers of this book may be somewhat surprised to learn just how much we are already capable of technologically, since progress is made at such a remarkable pace. So remarkable, in fact, that the vast majority of this progress is met with little fanfare, press coverage, or public announcements. As a result, the general public is largely left in the dark about what we, as a society, are already capable of achieving using technologies available today.

As we move forward, and the rate of technological advancement continues to increase, an understanding of where we are technologically will become crucial to understanding where we're headed. This book's goal is to provide a foundation of knowledge

13

about some of the most interesting technological advancements in the realm of entertainment.

Let's begin our journey by taking a look at the somewhat symbiotic relationship that has developed between the technology and entertainment industries. To keep things simple, let's drill down to one section of the entertainment industry and only concern ourselves with the recent trends in the motion picture industry.

Nearly every year, new movies raise the bar for what's considered state-of-the-art special effects. These movies consistently impress, and audience attendance rates have continued to climb -- even in a weak economy.[2] Meanwhile, industries related to consumer electronics and information technologies have also continued to thrive by offering devices and services that allow consumers to enjoy their media in an ever-growing number of ways. Demand for these gadgets, in turn, has pushed technological innovation to new levels. Batteries last longer, screens have increasingly better resolution, and user interfaces have become more intuitive; just to

---

[2] "Theatrical Market Statistics." *MPAA*. N.p., n.d. Web. 31 May 2014. <http://www.mpaa.org/wp-content/uploads/2014/03/MPAA-Theatrical-Market-Statistics-2013_032514-v2.pdf>.

name a few. Innovation in one arena fuels innovation in the other. However, this symbiotic relationship is not limited to the motion picture industry. In fact, as we will see, a similar relationship can be found in nearly every corner of the entertainment industry.

This book offers a guided tour of where technologies used for entertainment purposes are at today, as well as where they are poised to move towards in the coming years. While the term "entertainment" is rather encompassing, this text will primarily focus on movies, television, music, video games, and printed media. It should further be noted that although there are numerous directions that any of these given sectors could move towards in the future, it is simply not realistic to try to cover every possibility. Therefore, the main focus of this book will be on some of the most interesting paths that these technologies could take. With this in mind, our aim is to strike a balance between interesting and plausible. That is, it would be *interesting* if in the future we could turn on an old TV show and have a conversation with the actors, but it really is not very *plausible* to imagine a technology that would ever offer that kind of capability.[3] Likewise, it is almost guaranteed that we will have higher definition television sets in the coming years, but that just isn't very *interesting*. Again, a balance of these two extremes is the overriding goal.

First, we'll take a look at what types of innovations are currently under development, followed by an exploration of some of the more incredible, yet still possible, paths that we may see unfold. We begin our tour, however, with a look at some of the most profound ways in which technology has impacted the entertainment industry in recent years.

---

[3] While it's true that you could certainly simulate a conversation with an actor, the words wouldn't really be theirs. Rather, they would be the result of an algorithm or third-party source.

Chapter 2: Technology's Role in Entertainment

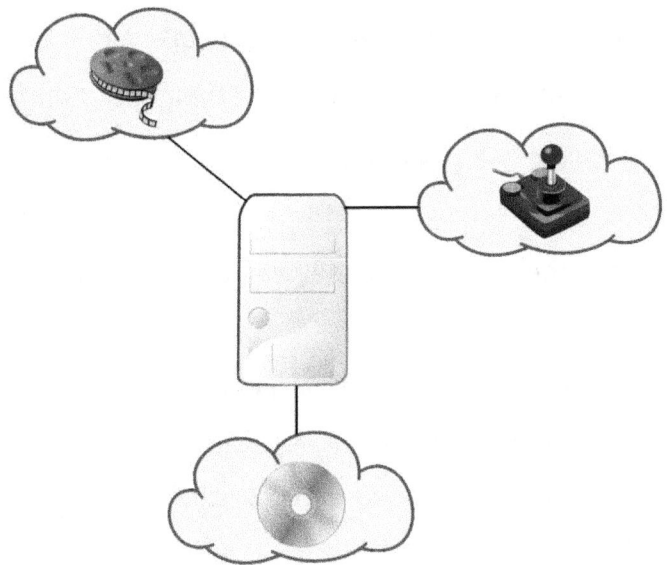

A Brief History of Entertainment Tech

The entertainment industry is in the middle of one of the largest and most exciting transformations that it's ever faced. As you would likely suspect, this transition is centered on recent technological innovations. To truly appreciate just how large of a transformation we're experiencing, it is imperative that we take a look at how most people obtained and enjoyed their entertainment media just a few years ago.

If someone wanted to watch a newly released movie in the early 1990s, they had the following options:

1. Go to the movie theater or drive-in.
2. Wait for the movie to be released on VHS.
3. Wait (even longer) for the movie to be broadcast on television.

In 1995, the widespread adoption of the DVD resulted in a new option that offered an increase in picture quality when watching movies at home. Nevertheless, the choices available to the average consumer were still quite limited.

As the years went on, this structured movie-watching process started to undergo a transformation. Not only were more media formats introduced, but we also began to see more ways of obtaining and viewing our media. The standard media format, VHS, was eventually replaced by DVD, which is currently being replaced by Blu-Ray. Televisions, the dominant entertainment screen in the household for over 50 years, have begun to cede their foothold to a variety of other small screen devices including laptops, cell-phones, and tablet computers.

The most profound transformation, however, revolves around the mainstream's discovery that the Internet is completely capable of delivering high-quality media to the home in a quick and convenient fashion. The catalyst of this transformation, in many ways, was the same as what spurred change in numerous other media industries: *Napster.*

For the uninitiated, Napster was the first widely adopted application that allowed users to "share" music and other media with one another using the Internet. At first blush, the software may not seem like the most obvious change agent that altered the way consumers enjoyed movies. After all, Napster is usually only

19

mentioned when discussing the music industry. While it is true that music was the primary commodity traded on Napster's networks during its short peak of popularity between 2000 and 2001, its impact influenced consumer perspective and behavior in nearly every sector of the entertainment industry. Napster taught a generation that media did not have to come from traditional sources and that the rules were changing. In addition, it introduced the general public to the idea that there were easier and more convenient ways to enjoy the media which they desired.

By late 2001, there were numerous file-sharing networks a tech-savvy consumer could choose from to find nearly any movie, album, book, or software-application he or she desired. This marked the first time an average person could locate, transfer, and consume their media using the Internet and their home computer. One of the largest transformations ever experienced by the media industry was underway. Still, there were a few equally critical transitions that did not occur until more recently.

One industry-changing transition, which can be traced back to a few years before the invention of

Napster, was the introduction of the wildly successful business known as Netflix.[4]

The Rise of Internet Video

This California-based company was originally dedicated to delivering DVDs by mail, but has since shifted its focus to delivering high-quality video content over the Internet. Since its inception, Netflix has turned the video distribution industry upside down, *twice*. First, it took on industry giant Blockbuster in the DVD distribution market and came out on top by nearly all accounts, with Blockbuster finally filing for Chapter 11 bankruptcy protection in September 2010. This was significant because it not only solidified the notion that consumers were interested in easier ways of getting their media, but it also demonstrated that the industry was changing and businesses would either have to adapt or risk falling by the wayside.

The second game-changing maneuver carried out by Netflix was the introduction of "instant-streaming

---

[4] Netflix was born in 1997, to be exact.

"BookRenter Adds Netflix Co-Founder Marc Randolph To Its Board Of Directors." *TechCrunch*. N.p., n.d. Web. 14 Mar. 2014. <http://techcrunch.com/2010/05/07/marc-randolph-bookrenter/>.

video" for movies over the Internet. This service was rolled-out to a much broader audience in 2008, but was initially greeted with little fanfare. Since its release, however, this service has become enormously popular and has spurred other companies to create competing services of their own. The most popular of these include Amazon Prime Instant Video, Hulu, Vudu, and Crackle.

This event was important to the entertainment industry because it paved an entirely new path from the average consumer to their favorite movies. No longer was it necessary to visit the store or wait for a physical copy of the movie to show up in the mail. Instead, with an Internet connection and a computer, one could enjoy their chosen movie in a matter of seconds without ever leaving the house. What's more, this service was offered legitimately, with legal rights to the videos being served up. This meant that Netflix held one sure-fire advantage over the many file-sharing networks that had sprung up in the time since Napster; it was completely legal. The widespread acceptance of media consumption via the Internet had taken another giant step forward.

Mobile Takes Over

Not only has there been a change in how consumers get ahold of films and TV shows, but we're also starting to see a shift in how they watch such content. That is, rather than the television set serving as the sole provider of movies and television programs, desktop computers, laptops, tablets, and even smartphones have recently positioned themselves as viable alternatives.

This migration of content from the family television to a variety of smaller screens is meaningful because it offers mobility and convenience that traditional content-distribution channels cannot match. The hardware, software, and infrastructure of the cable and satellite companies simply aren't geared towards delivering content to a smartphone, tablet, or computer. These traditional providers could certainly expand into this new mobile-oriented market (and they are!), but the fact remains that the traditional methods of video distribution are not designed to compete in this arena.

To illustrate this point, consider how you would go about connecting your satellite dish to your tablet or connect to your home's cable TV line while out of town.[5] Traditional cable and satellite providers just cannot match the convenience of media delivered to mobile devices via the Internet. In short, these two newer areas of technological development will almost certainly have an enormous impact on the way consumers enjoy content in the coming years.

The Ripple Effect

The movie and television industries aren't the only ones undergoing some pretty major changes. In fact, the rise of the Internet as a content distribution system has had a major impact on nearly every section of the entertainment industry.

Consider the publishing industry as an example. Only a few short years ago, there was a clear path to get a book published and distributed to the general public (i.e. through a publisher). Today, however, the Internet has dramatically altered this age-old process so that the dream of publishing a book and getting it distributed on international channels can become a reality for anyone

---

[5] It should be noted that some cable providers are now providing mobile apps that allow subscribers to watch content while away from home. However, these apps don't take advantage of the traditional cable TV infrastructure and instead rely on the Internet to deliver content.

who sets their mind to it.[6]

The modern music industry presents a situation that is nearly identical. Up until recently, there was more or less a single way to get your music distributed to the masses: record labels. Once again, the Internet (and a home computer loaded with the appropriate software) has antiquated this traditional distribution system and paved the way for aspiring artists to release their music online for fans worldwide to enjoy.

One standout example of this new type of music distribution is the hip-hop duo from Seattle, Washington: Macklemore & Ryan Lewis.

Without a record label or recording contract, the duo made history in January of 2013 when their song "Thrift Shop" reached the number one spot on Billboard's Hot 100 charts, marking only the second time that such a feat had been accomplished.[7]

It's noteworthy that this breakout success wasn't

---

[6] One popular way to accomplish this is through Amazon's self-publishing service.

[7] Feeney, Nolan. "Macklemore's "Thrift Shop" Tops Billboard Chart Independently | TIME.com." *NewsFeed | Breaking news and updates from Time.com. News pictures, video, Twitter trends. | TIME.com*. N.p., n.d. Web. 18 Apr. 2013. <http://newsfeed.time.com/2013/01/25/macklemores-thrift-shop-is-first-indie-hit-to-top-charts-in-nearly-two-decades/>.

a result of a marketing push by record labels and major distribution channels, but was instead spurred in large part by a strong online presence and a relentless touring schedule.

Conventional distribution systems (i.e. record labels, radio, and physical media) haven't quite vanished, but as Macklemore and Ryan Lewis have proven, the significance of these major labels is beginning to come into question as new music-oriented distribution channels, services, and devices are introduced on a near-constant basis.

With all of this in mind, it's fairly safe to say that standard practices involving media content distribution and consumer consumption habits are venturing into uncharted waters. There is more media content available today than ever before, with more ways to enjoy this content, as well. Musical artists, filmmakers, and media companies have ample choices when it comes to how their works should be distributed. Meanwhile, consumers can decide exactly when, where, and how they'll experience their favorite music, movie, or television show. It really is astounding how many options there are for everyone involved.

Negative Impacts

Having all of these options is not *strictly* a positive move for consumers, however. Without a doubt, it's great to visit YouTube, Vimeo, or UStream and have millions of videos to choose from at that very moment, but this enormous amount of choice comes with some noticeable drawbacks. Mainly, such content distribution networks provide consumers with an "overload" of choices and a loss of a quality filter.

Although we'll save the in-depth discussion of internet-based video services for a later chapter, it's important to point out that today's consumers have far more entertainment choices than previous generations. In fact, there are so many options that it can sometimes be quite a challenge just to decide which song to listen to, which TV show to watch, or even which content-delivery service to use. While this is clearly not an extraordinary downside, it is nonetheless worth noting how more options for consumers lead to more complexity in such decisions.[8] In addition, this

---

[8] Anyone who has ever spent more time picking out a video to watch on Netflix than the time it took to watch that video will surely be able to relate.

overwhelming number of choices also comes with a second concern: a diminished content filter.

A few years ago, most people relied upon a few main channels of content distribution such as television networks, large movie studios, record labels, and major book publishers to provide them with their entertainment. These traditional pipelines had their downsides but at the end of the day, they were interested in making quality entertainment that would bring in the largest audience and the most money.

Today, these traditional distribution channels still exist, but are in some ways beginning to be overshadowed by the sheer volume of modern Internet-based video outlets. Without a doubt, this increase in choice is a big win for consumers. The tradeoff here, however, is a degradation of the "quality filter" that accompanies the more traditional video delivery methods.

In other words, a global release of a truly terrible movie is *somewhat* of a rarity because releasing a bad film hurts a studio's wallet. Compare that to the type of content uploaded to popular video-sharing sites, like YouTube or Vimeo, and the notion that we are losing our "quality filter" starts to become clearer. These amateur-based video websites provide a great deal of entertaining content, but they also house hours upon hours of footage that very few people would deem worth watching. Finding a worthwhile video is certainly possible, but it requires a bit more time and effort to sift through the undesirable content first. In this respect, professional content-creation industries still have an advantage over these amateur-driven entertainment networks.

Note that I'm not implying professionals only make good movies - they most certainly do not. Nor am I implying that it is impossible to find a good video on these amateur-driven networks - it is not. However, the *ratio* of quality content to utterly awful content is much higher for professionally made films than it is on the amateur-oriented video websites. At the end of the day, all of this results in a more difficult decision for today's consumer since, not only are there more choices to consider, but many of those choices will almost certainly leave them dissatisfied.

Consumers of entertainment media aren't the only ones feeling some negative impact from Internet-based content distribution. In fact, content *creators* may actually experience more drawbacks than consumers. First, consider the fact that there are approximately 8 years (70,120 hours) of content uploaded to YouTube *every day*.[9] This means that if you're an aspiring actor, movie producer, or short-film creator, and you decide to upload your 1-hour long video to YouTube, then your video is only about $1/_{70, 120}$ of all of the content that was uploaded that day.

It should be noted that these statistics are only taking into consideration *one* of the countless websites that exist to host and deliver web-based videos. Even if we narrow this number down to a few of the most popular video-sharing sites (YouTube, Vimeo, YouKu, and Flickr Video), we're still talking about hundreds of millions of hours of video content available online competing with your footage. Of course, this doesn't mean that all content has an equal chance of being viewed, since some videos are clearly more appealing than others. Nevertheless, it does illustrate just how much content is out there and provide a gauge of how likely (or unlikely) it is that your video will become the

---

[9] "Press room - YouTube." *YouTube*. N.p., n.d. Web. 18 Apr. 2013. <http://www.youtube.com/t/faq>.

next "big thing."

To further delineate just how much the odds are stacked against an aspiring artist, consider the sizes of some of the largest online music retailers in the world today:

- *Pandora* - over 900,000 songs[10]
- *Rhapsody* - over 16 million songs[11]
- *Amazon MP3* - over 16 million songs [12]
- *Apple iTunes* - over 26 million songs[13]

Again, however, these are just a selection of some of the most popular music sites that had data available at the time of writing. There are countless other options available and perhaps even more songs available on those services not listed here. However, even from such a small sample of music distributors, it's clear that by creating a path for anyone to release their content to the

---

[10] "Music Services Compared: Pandora, Spotify, Slacker and iHeartRadio" *Mashable*. N.p., 13 Feb. 2013. Web. 10 May 2014. <http://mashable.com/2013/02/13/music-services-compared-2/>.

[11] "Discover Rhapsody - What is Rhapsody?." *Rhapsody:: Subscription Music Service: Listen All You Want: Millions of Songs*. N.p., n.d. Web. 8 Apr. 2013. <http://www.rhapsody.com/about/index.html>.

[12] "Amazon MP3." *Wikipedia, the free encyclopedia*. N.p., n.d. Web. 7 Sept. 2013. <http://en.wikipedia.org/wiki/Amazon_MPP3&oldid=555350180>.

[13] "Apple Unveils New iTunes." *Apple*. N.p., n.d. Web. 7 Sept. 2013. <http://www.apple.com/pr/library/2012/09/12Apple-Unveils-New-iTunes.html>.

world, the Internet has also facilitated levels of competition for content creators that have never been seen before. In short, although the Internet is undoubtedly the greatest content distribution network the world has ever seen, it has permanently changed the processes of content creation, distribution, and selection.

In the coming years, anyone involved with any type of media creation or distribution will certainly have to consider what impact the Internet will have on their work. The Internet has created stiff competition, provided potential-customers with a multitude of choices, and perhaps forever altered the way in which consumers choose which movie, TV show, or song they will enjoy next. Disregarding the impact of the Internet on consumer choice will almost certainly leave content creators and distributors vulnerable to becoming obsolete, as the most skilled competition will embrace and capitalize on these changing trends.

It is worth mentioning that the entertainment industries have felt another major impact from the Internet revolution: unauthorized downloading of copyrighted content (i.e. media piracy). There is considerable debate about how much of an impact this type of behavior has had on the industry in terms of profits and sustainability. However, there is much less debate surrounding the idea that the response by most corporations (i.e. million-dollar lawsuits brought against consumers) has (1) not remedied the problem, and (2) actually damaged the relationship between content creators, such as movie studios, and customers.

With these two ideas in mind, it seems likely we may see less lawsuits brought against consumers in the future, and more innovative strategies to curtail copyright infringement. However, this is far from a sure thing as there are numerous considerations in this area. We will take a closer look at this area in the next chapter.

Multi-Purpose Devices

While the Internet is the technological innovation that has arguably had the largest impact on the entertainment industry, it is far from the only such innovation. Indeed, the introduction and widespread use of a variety of mobile devices has resulted in a situation where, in recent years, consumers have owned many devices that only really excel in one (or a few) particular area(s). For example, video game consoles were primarily used to play video games. Cell phones were used to make phone calls, and send text messages to friends. Desktop computers were used for more heavy-duty tasks like web development and computation-heavy calculations. Of course, these are generalities and there are exceptions to the rule, but by and large, most devices up until recent years were made for some primary purpose.

In the past few years, however, there has been a shift in the direction of technology and in the mindset of consumers. We have started to see video game consoles branch out and become general entertainment devices offering the capabilities to play DVDs, stream Netflix, play music, and browse the Internet. For example, the recently-released Xbox One continues to blur the lines of what a video game console does by integrating the ability to watch live TV. Cell phones have become "smart" and now allow users to browse the Internet, play games, view and edit documents, stream music, and much more.

Even the emergence of tablets is a testament to this movement. Slates offer capabilities somewhere between a smartphone and a desktop computer; compact enough to be mobile and powerful enough to handle intensive computations.

With this in mind, all of the signs point to a future where technological devices serve a multitude of purposes. The distinctions between various devices will become less pronounced. Mobile devices will become more powerful, powerful devices will become more compact, and the days of consumers owning many different single-purpose devices will start to fade. Instead, it is likely that we will see one compact gadget, powerful enough to handle even the most computation-intensive tasks, become the centerpiece of the average consumer's entertainment, business, and technological universes.

Although at this point it is anyone's guess as to what these types of devices will ultimately look like or how they will operate, it seems extraordinarily probable that we *will* see this trend continue, with more and more devices expanding upon their current capabilities. Eventually, it may be possible to have one device seamlessly replace the responsibilities currently undertaken by nearly every other entertainment device in the home. Indeed, mobile computing has already started to eat into the popularity of desktop computers, and it may only be a matter of time until other categories of devices are affected similarly.

Aside from the possibilities discussed in this section, there are numerous other trends that have emerged in recent years and have had an effect on the entertainment industry as a whole. Now that we've taken a broad look, we turn our attention to how individual areas of the entertainment industry have been affected by technology in recent years, which trends have emerged in each, and where these individual sub-industries may be headed.

Discussion Questions:

1) Do you agree with the author's assertion that the entertainment industry is in the middle of one of the largest and most exciting transformations that have ever occurred? Why or why not?

2) Do you believe that file-sharing software (such as Napster) had a positive impact on the entertainment industry? Explain.

3) Has the increased amount of video, music, and other entertainment choices had a greater positive impact on content consumers *or* content creators? Explain.

4) Are the increased amount of content and delivery options good for the entertainment industry as a whole? Why or why not?

5) Do you agree with author's vision of a future where numerous technological devices available today converge into one all-powerful device? If so, describe the capabilities of that gadget. If not, explain what types of devices you envision being popular in the future.

Chapter 3: Movies

The film industry's use of technology, by and large, has not been extraordinarily innovative in recent years. This isn't to say that it doesn't build upon technological innovation to produce some very exciting, exhilarating, and breathtaking effects on screen. However, the business as a whole just hasn't pushed many boundaries in terms of the typical movie-going experience.

Thankfully, there are a few notable (and exciting!) exceptions to the otherwise lackluster rate of film industry innovation. As you might expect, most of these advancements are occurring on a small scale and have not yet hit mainstream Hollywood. However, there are a few that have started to catch on with the general public. There is one innovation, in particular, that many have been claiming will become the "next big thing" for over 75 years, which has finally started to gain some momentum with the general public: 3D movies.

3D Films

Most readers are probably familiar with the concept of a 3D movie. Most are also likely well-aware of the fact that it's become popular to offer moviegoers the option of enjoying films in either 2D or 3D. Some folks may be surprised to learn, however, that 3D motion pictures are not necessarily cutting-edge technology. In fact, primitive 3D films can be traced all the way back to 1922, which saw the release of the "Power of Love," a film widely credited as the first 3D movie.[14] It was not received extraordinarily well by the public, and 3D movies were relegated to the fringes of the budding film industry. Nevertheless, the idea of a 3D movie had been put into practice over 90 years ago -- hardly a cutting-edge concept, indeed.

Since the 1920s, there have been numerous points in history where it seemed like 3D may finally become the primary way to enjoy a movie. Notable 3D films can be found in nearly every decade since the 1950s: *Bwana Devil*, *Man in the Dark*, *House of Wax* (1950s), *The Mask* (1960s), *The Stewardesses* (1970s),

---

[14] "The Power of Love." *IMDb*. IMDb.com, n.d. Web. 15 Mar. 2014. <http://www.imdb.com/title/tt0013506/>.

[14] "A Tour Through the History of 3-D Movies." *REELZ*. N.p., n.d. Web. 15 Mar. 2014. <http://www.reelz.com/article/816/a-tour-through-the-history-of-3-d-movies/>.

[14] "The History of 3D Movie Tech." *IGN*. N.p., n.d. Web. 15 Mar. 2014. <http://www.ign.com/articles/2010/04/23/the-history-of-3d-movie-tech>.

Captain Eo (1980s), and *Echoes of the Sun* (1990s).[15]
However, to say that 3D had gone "mainstream" during
any of these time periods would be misleading. For all
intents and purposes, 3D was still very much the
exception to the rule up until the release of *Avatar* in
2009.

Since *Avatar*'s release, there has been an
enormous surge in the production of 3D movies.
However, it is worth pointing out that in recent months
(early Summer 2013) there has been a notable decline in
the popularity of 3D and the question is once again being
asked whether or not it will ever become ingrained in the
movie-going experience.[16]

Of course, only time can ultimately answer this
question. Nevertheless, it seems worthwhile to look at
the history of 3D, understand why the general public
appears to be resistant to the tech, and use this
information as a starting point in determining where

---

[15] "The History of 3D Movie Tech." *IGN*. N.p., n.d. Web. 15 Mar. 2014.
<http://www.ign.com/articles/2010/04/23/the-history-of-3d-movie-tech>.

[16] "3D films losing their appeal as UK ticket sales slump by a third | Mail
Online."*Daily Mail*. N.p., n.d. Web. 7 Sept. 2013.
<http://www.dailymail.co.uk/news/article-2319249/3D-films-losing-appeal-
UK-ticket-sales-slump-third.html>.

we're headed.

The main argument as to why 3D will eventually catch on in a permanent manner is that it seems like the logical next step in the movie-going experience. That is, humans are three-dimensional beings and we experience the world as such. It only makes sense that our entertainment should utilize all of the dimensions we are capable of perceiving now that it's technologically possible. However, it should be noted that this same line of reasoning was used way back in the 1920's as the basis for the prediction that 3D entertainment was just on the horizon.[17]

The counterargument is equally as intuitive. Since 3D has been the predicted the movie industry's "next big thing" for over 90 years, it's about time we realize it is never going to become the permanent standard. Proponents of this argument will point out that 3D motion pictures can cause eye-strain, headaches, and fatigue. Furthermore, if glasses are required, they can be quite uncomfortable and present challenges to people

---

[17] "Glasses-Free 3D and Smell-o-Vision: Movies of the Future from 1935."*Paleofuture*. N.p., n.d. Web. 7 Sept. 2013. <http://paleofuture.gizmodo.com/glasses-free-3d-and-smell-o-vision-movies-of-the-futur-513784216>.

who already wear prescription eyeglasses. Each of these items may not seem extraordinarily significant by themselves, but taken as a whole, the case begins to be made that 3D films aren't as practical or convenient as traditional cinema.

At this point, it's anyone's guess as to which of these arguments is correct. Based on history alone, however, it seems highly likely that even if 3D movies once again drift away from the mainstream, they will always have a place at the fringes of the film industry.

4D Films

A noteworthy offshoot of the 3D film craze is something that's come to be known as the "4D movie." Reportedly first appearing in the mid-1980s, the 4D flick uses moving seats, fans, odors, mist, water droplets, flashing lights and other physical effects in combination with a 3D film to create a unique immersive entertainment experience.

For example, during one scene in *Transformers: The Ride*, the physical effects of an explosion are imitated via hot air being blown while fog rises around the viewers. In *Marvel Super Heroes 4D*, the viewer

43

experiences "air jets at floor and head level from all directions, water sprays, and an unexpected claw-jab through the seat back from Wolverine...". [18]

Unsurprisingly, the availability of these films is significantly less than that of 3D movies; largely due to the extra equipment and accompanying expenses necessary to create the experience. As a result, it's likely that these types of films won't replace traditional cinema anytime soon. Nevertheless, production of these movies has picked up speed in recent years and it seems that 4D movies are not going away; at least not in the near future.

The concept of the 4D movie is interesting not only because it offers viewers a novel experience unmatched by traditional cinema, but also because it blurs the lines between passive and interactive entertainment. That is, rather than merely watching a story unfold around them, an audience is instead thrust into the middle of the action where they experience some of the same sights, smells, and physical effects as

[18] Review: Madame Tussauds' Marvel Superheroes 4D Exhibition." *Bleeding Cool Comic Book, Movies and TV News and Rumors*. N.p., n.d. Web. 11 May 2013. <http://www.bleedingcool.com/2010/06/03/review-madame-tussauds-marvel-superheroes-4d-exhibition>.

the characters on screen. In fact, in some rare instances, audiences have actually been given the ability to help determine the direction that a story will take during the movie. Of course, this crosses into territory that some would classify as a video game. Regardless of the classification, however, this blend of 4D and interactive entertainment does offer quite an interesting medium to tell a story. It will certainly be interesting to see if and how these elements are combined to bring about new experiences in the years to come.

Second Screen Technologies in Movies

Yet another exciting area of interactive entertainment that we may see movies branch out to is something known as the "second screen" or "social entertainment." The second screen moniker describes the use of a smartphone or tablet to enhance the media shown on the primary screen. In recent years, this type of entertainment has really started to take off in the realm of television. As such, we will postpone an in-depth look at this technology until we reach that chapter. However, it's worthwhile to mention that the somewhat controversial use of the second screen in movie theaters officially made its debut in April 2013, with the release of a film titled "App" in the Netherlands.[19] The movie is, somewhat fittingly, centered on how technology is changing our lives.

Throughout the film, there are 35 occasions where bonus content is displayed on the viewer's second screen, designed to enhance the experience.

At one point in the film when a few characters are texting each other, the content of the text messages is displayed on the viewer's mobile device. Audience members who choose to view this extra content may have a slightly more engaging movie-going experience. Meanwhile, since the text message content is not crucial to the storyline, those who choose not to participate can still enjoy the film. Note that the potential here goes far beyond this rather basic use of the technology, but we will explore more of these possibilities a bit later.

Before venturing into that area, however, we'll focus our attention on innovations surrounding another aspect of the movie consumer's experience: movie

---

[19] "The Second Screen Comes To The Movies With App-Enhanced Film, "App"."*Co.Create | creativity + culture + commerce.* N.p., n.d. Web. 11 May 2013. <http://www.fastcocreate.com/1682579/the-second-screen-comes-to-the-movies-with-app-enhanced-film-app>.

distribution.

Movie Distribution

The term "movie distribution" doesn't typically conjure thoughts of excitement or wonder. Nevertheless, there are actually very few other areas of the motion picture industry that offer so much room for innovation.

To be clear, movie distribution really is just the term used to describe how a person obtains a film. By this point in the book, readers can probably guess that the Internet and personal handheld devices both play central roles in this area of potential innovation.

More specifically, the rise and proliferation of the Internet has resulted in the most efficient and impressive content delivery systems the world has ever seen. This foundation has allowed services like Netflix, Vudu, and Amazon Instant Video to spring-up, flourish, and thrive. As these types of businesses grow in popularity and branch out to more countries, opportunities are created for the film industry to reach an untapped base of customers.

Similarly, as powerful mobile devices such as smartphones and tablets become the norm worldwide, opportunities are created for the film industry to reach more patrons. Not everyone can afford to splurge on an internet-connected television or Blu-ray player, but justifying a mobile device is a little easier thanks to its ability to accomplish a wide variety of tasks. Since most modern smartphones and tablets are more than capable of playing a variety of video types, these devices open the door to a world of on-demand entertainment that just wouldn't exist otherwise.

Looking into the future, we find the potential for online movies to be released in a standardized universal format across the industry. That is, regardless of which device you own, which movie you'd like to watch, or which company that movie is purchased from, it'll be available to download and enjoy without hassle. This type of uniformity was enjoyed by consumers before the rise of digital media (e.g. VHS and DVD), and would likely be just as beneficial to consumers in this new digital era. If the film titans can come together and make this a reality, it would not only benefit the consumers, but the industry as a whole.

Whether we'll ever actually see this occur, however, is an uncertainty, at best. Nevertheless, no technological limit stands in the way of such an innovation. Yet, it seems there may be too many hurdles in place for this vision to ever be fully realized. I've been pleasantly surprised by similarly unlikely events in the past, though.

To conclude our tour through the world of cinema, we turn to a recent development in the industry that was spurred by technology, yet many in the industry wish had never arisen: online movie piracy.

Movie Piracy

One of the largest impacts of Napster and file sharing was a shift in how many consumers obtained films. While it's fairly safe to say that no entertainment business looks favorably upon consumers downloading copyrighted content free of charge, the motion picture industry's response has been widely viewed as particularly harsh.

In 2010, for example, the movie studio behind the Oscar-winning film, *The Hurt Locker*, brought forth a copyright infringement lawsuit onto over 24,000

individuals accused of illegally downloading the film.[20]
The studio reportedly required each of the defendants to
pay between $1,000 and $3,000 to settle out of court and
avoid the possibility of financial ruin.[21] Using the average
of these numbers ($2,000), some quick math reveals that
the studio was hoping to bring in $48 million from this
suit. To put this in perspective, the film brought in just
over $49 million at the box office worldwide. This was,
without a doubt, the largest and most ambitious lawsuit
ever brought against alleged downloaders of copyrighted
materials to date.

The lawsuit made headlines around the world and
sparked cries of outrage by many, including numerous
individuals who felt that their name had been
erroneously included in the list of alleged infringers.
Stories like these, combined with the large number of
individuals targeted and the high amounts requested

---

[20] "Hurt Locker Piracy Lawsuit Abandoned, Court Records Show." *Huffington Post Canada - Canadian News Stories, Breaking News, Opinion*. N.p., n.d. Web. 22 June 2013. <http://www.huffingtonpost.ca/2012/03/30/hurt-locker-piracy-lawsuit-abandoned_n_1392427.html>.

[21] Purewal, Sarah Jacobsson. "'Hurt Locker' Lawsuit Targets 24,583 BitTorrent Users."*PCWorld - News, tips and reviews from the experts on PCs, Windows, and more*. N.p., n.d. Web. 7 Sept. 2013. <http://www.pcworld.com/article/228519/Hurt_Locker_Lawsuit_Targets_24583_BitTorrent_Users.html>.

resulted in the suit being viewed in a particularly unfavorable light by the general public. It is noteworthy that the lawsuit was eventually dropped in 2012, but not before an undisclosed number of individuals agreed to the settlement.

The use of these tactics as a means to deter illegal downloading is not limited to the studio responsible for *The Hurt Locker.* In fact, this type of legal action isn't even solely used for the illegal download of mainstream blockbusters. In 2011, a lawsuit was brought against those whom allegedly downloaded a B-movie titled *Nude Nuns With Big Guns.* Interestingly enough, the suit may have been part of the profit-earning strategy from the beginning. According to at least one Seattle-based attorney, Lory Lybeck, the *Nuns* lawsuit is nothing more than a "mass copyright litigation machine."[22] According to Lybeck, most people are inclined to settle when threatened with such legal action because they don't want it to become public knowledge that they allegedly downloaded movies with provocative or sexually-charged

---

[22] "How Mass BitTorrent Lawsuits Turn Low-Budget Movies Into Big Bucks | Threat Level." *wired.com* . N.p., n.d. Web. 7 Sept. 2013. <http://www.wired.com/threatlevel/2011/03/bittorrent/>.

titles. It's a strategy that's certainly earned movie studios some additional profits in recent years, but the long-term costs of such tactics are difficult to estimate.

As you can imagine, these types of tactics have had mixed results. Hearing of such lawsuits has almost certainly deterred at least some individuals from engaging in copyright-infringing behavior. This may have reduced the number of folks illegally downloading movies. However, there is no guarantee that these individuals have instead legitimately purchased the movie. Thus, it's hard to say whether the "deterrent effect" of these lawsuits actually generate additional revenue for movie studios.

Furthermore, it's important to highlight that these types of suits don't facilitate a positive relationship between the film industry and the general public. It's even possible that these lawsuits cause some people to avoid purchases that would fund the film industry and their lawsuit efforts.

Taking all of this into account, it seems that there are a few likely possibilities when determining how the rise of digital piracy and the resulting litigation may affect the industry in the future.

First, movie studios may continue following in the footsteps of *Nuns*, and utilize lawsuits as a way to offset losses suffered as a result of piracy. The main risk here is that this is a fairly unpopular tactic, which risks alienating the film industry from the general public. This may ultimately build disdain and resentment towards the industry, which could lead to higher rates of piracy. In other words, while this type of strategy may lead to a better bottom line in the short term, it's entirely possible that long-term repercussions may outweigh those benefits.

A second direction the film industry could take is to cut back on suits after realizing that they damage public relations. The industry could launch efforts to build goodwill and win support from the general public. A crucial part of this plan involves making it less socially acceptable to download movies. It is fairly well known that media piracy, as a whole, is particularly socially acceptable for an illegal activity. In fact, a Danish study recently found that 70 percent of the general public believed that downloading media for personal use was socially acceptable. If the industry can find a way to reverse this trend then there's a real chance that rates of

53

piracy will decline without resorting to lawsuits. Of course, this is a rather optimistic view of how events may unfold in the future and, as such, it's not extraordinarily likely.

A third possibility is that we'll see stronger legislation to curb the trend of media piracy. In recent years, we've already seen two very high-profile attempts to pass such legislation: the *Stop Online Piracy Act* (SOPA) and *PROTECT IP Act* (PIPA).[23]

The goals of these proposed laws were to reduce the availability of pirated media online by increasing the government's ability to shut down websites and servers believed to host and distribute copyrighted materials illegally. Neither of these bills passed, largely due to public outcry over privacy and anonymity concerns. Nevertheless, the industry has seemingly tipped its hand as to how it intends on dealing with this issue in the future.

Going forward, it seems highly probable that

---

[23] Magid, Larry. "What Are SOPA and PIPA And Why All The Fuss?." *Forbes*. Forbes Magazine, 18 Jan. 2012. Web. 14 Mar. 2014.
<http://www.forbes.com/sites/larrymagid/2012/01/18/what-are-sopa-and-pipa-and-why-all-the-fuss/>.

23

there will be similar legislation signed into law that will primarily aid the film industry in their battle against piracy. The overall effectiveness of such an action is far more unpredictable, however, since there's no telling how the public may react to such legislation. It is possible that people may respond by finding loopholes, weak points, or workarounds to the legislation and continue their piracy without breaking stride. Still, others may increase their piracy efforts as a way of protesting the law. Others may capitalize on the law by downloading and selling pirated materials to others who are too fearful to continue downloading. How all of these factors come together will determine the success of movie-industry-backed piracy legislation, should the industry choose to continue moving down this path in the future.

The film industry certainly has some interesting technological innovations occurring today, as well as on the horizon. While one section is busy utilizing novel techniques to provide new and exciting immersive entertainment experiences, another is well poised to enable the delivery of content to customers in a faster and more convenient fashion. Of course, possibly the most interesting development to watch for is how the

55

industry adapts to the changes resulting from the rise of the Internet and Internet piracy. However this last item pans out, I suspect its development will be interesting to watch unfold.

Discussion Questions:

1) Have 3D movies finally become a permanent fixture in mainstream culture or are we simply in the midst of another fad? What trends in the industry support your opinion? Explain.

2) If you heard that there was bonus content that would be shown on second screen devices throughout a movie, what effect would that have on your decision to see the film? Would this technology make you more interested? Less interested? Neither? Explain.

3) Will streaming movies over the Internet one day be replaced by a newer technology? If so, what advantages would such a futuristic technology need to have in order to be considered superior? If not, describe how you see this technology improving and evolving in the years ahead.

4) Are those who download movies without paying for them hurting the film industry? Explain. (Be sure to include whether you believe the film industry will be able to continue to be profitable if movie piracy continues.)

5) Should those who download movies without paying for them be punished? If so, what punishment do you consider suitable? If you don't think there should be a punishment, explain why not.

6) What do you see as the most likely action that will be taken to curb movie piracy in the future? How effective do you think this strategy will be? Explain.

Chapter 4: Television

The future of the television industry will undoubtedly experience many of the same challenges, innovations, and changes as those facing the movie business. Piracy will threatens the bottom line as more and more television shows are downloaded or streamed from the Internet. 3D TVs and 3D programming have started to appear in recent years, but have struggled to catch on in any meaningful way. Perhaps the most impactful trend that we've seen recently is the shift towards "social television."

Social Television

The use of a social component to enhance broadcasts has been rapidly spreading across the industry. Online chatter about shows has been catching on with creators and viewers alike, which has resulted in "social television" becoming the frontrunner as the next major advancement in the realm of television. Thus, we begin this section with an exploration of what social television truly means and how this innovation may ultimately impact the television industry in the years ahead.

Social television revolves around the idea that

people like to communicate with each other about the television programs they're watching. Regardless of whether they're enjoying the latest television hit, a large sporting event, or a breaking news story, most people seem compelled to make interesting, insightful, humorous, or even general comments to one another while stories unfold in front of them. This particular aspect of social TV is nothing new.

For decades this commentary was typically confined to the family room, pub, or other gathering spot during the broadcast. Viewers would make quick remarks to one another during the airing of the show, discuss it in more detail during commercial breaks, and reflect on the program together after it had finished.[24]

In the past, even when someone watched a show alone, it was commonplace to discuss the events of a major broadcast the next day at the water cooler with co-workers, or to call up a friend during a particularly surprising, unexpected, or historical broadcast to make sure their friend had witnessed the same event, and then

---

[24] Anyone unfamiliar with this type of behavior need only visit a local tavern showing a popular sporting event. Find a group of fans watching the game together and observe.

discuss its importance. All of these activities dealing with socializing around television have already been happening for decades.

"Social TV," however, is the term that has been used more recently to describe how the use of technology will allow these types of social activities to occur in real-time between people across the globe, as well as create novel channels of dialogue such as direct conversations between the viewers and the creators of shows. In other words, the aim of Social TV is to bring social commentary to the digital realm and create experiences that are engaging, interactive, and unique.

Before continuing, it's worth noting that there's no standard definition of what qualifies television as "Social TV." There is no rubric or list of guidelines that must be followed in order to be classified as social. Nevertheless, the presence of any sort of social component is usually enough to warrant the classification. In recent times, the most popular way to transform a television show into a *social* television show has been through the use of the social networking site *Twitter*. However, most other major social networks have played some role in making a television show more social at one time or another.

If a program is using Twitter as a means of creating a social TV experience, the typical formula is to inform the viewer that they can tweet (send a short, publicly visible message) to a username owned by the show or one of its actors. In addition, Twitter also supports the use of hashtags (similar to a keyword) in messages to help users categorize their message and improve the chances that their viewpoint will be heard. Either way, if the tweet is seen by television show's producers, and it's deemed interesting, the message may ultimately be incorporated into the broadcast of the show.

Viewers can also use their laptop, tablet, or smartphone (usually just referred to as the "second screen") to see and create tweets related to the show in real-time. Since these tweets are public, they'll be visible to those in charge of the television show, as well as by other viewers. The creators of the program and other fans of the show can then respond. In this way, a conversation is started, and more "buzz" is created around the TV show. Again, the exact form of interaction is going to vary slightly depending on which social network is in use, but the formula and end goal is typically the same: start a conversation about the program to create a more engaging experience.

To many readers, this probably isn't a foreign concept since this type of activity has become fairly commonplace over the past few years. Numerous shows follow this basic formula to varying degrees and with varying levels of success. These days, however, just about every prime time show encourages viewers to check out their social media counterpart at some point during the broadcast. A few programs that have capitalized on their use of social media include *America's Got Talent*, *The Bachelorette,* and *So You Think You Can Dance* -- just to name a few.

Another big hit in terms of social media interaction has been the college-crowd favorite, *Workaholics.* For the most part, none of these shows have done anything revolutionary in terms of integrating social media. They simply have had solid programs with viewers anxious to discuss them with other fans and watch bonus content online.

There are a few television programs that have gone above and beyond the typical approach just described however, and have thus stood out as pioneers in the social TV arena. One such program was the 2011 *MTV Video Music Awards*.

Not only did MTV encourage viewers to tweet to the show during the live broadcast, but they were also offered a tool to see which artists were trending (being mentioned the most by other viewers) and which photos of the event were being viewed the most by others. In addition, MTV even featured a map of where artists were seated at the event that indicated what they were tweeting in real time. The event was truly groundbreaking in that it allowed viewers to not only watch the event live on television, but to also experience the event and interact with other fans, all without leaving home. It's no surprise that MTV's *Video Music Awards* went on to win a "Shorty Award" that year for their exemplary use of social media.[25]

Some television shows have offered a social

---

[25] According to the official *Shorty Industry Awards* website, the Shorty Awards honor the best brands, agencies, and professionals on social media.

"VMA Twitter Tracker." *Honoring the Industry's Best Agencies and Social Media Leaders - Shorty Industry Awards*. N.p., n.d. Web. 4 July 2013. <http://industry.shortyawards.com/category/4th_annual/live_television/e1/vma-twitter-tracker>.

component that not only encouraged viewers to discuss the events of the program, but influenced the show in some form or fashion. Typically, this type of interaction is offered by live broadcasts requesting viewers to vote for a winner or determine what they think of a particular issue. These votes are usually tallied and the results have an effect on some portion of the show. Some popular programs that offer such an experience include *American Idol*, *Dancing with The Stars*, *The Voice*, and *The X Factor*.

To be sure, voting for contestants and determining the outcome of a televised talent show is not, in itself, anything new. However, encouraging viewers to communicate with the show, with other viewers, and vote on the program's outcome -- all while simultaneously visiting links filled with bonus content related to the television program -- really encompasses nearly everything the term "social TV" represents today.

Although Twitter is currently the leading social media platform of choice for television networks interested in making their programs more "social," it's far from the only option. Other leading social media sites such as *Facebook*, *Google+,* and *Instagram* have all played a role in facilitating conversations between fans of shows and allowing viewers to have their thoughts heard by producers. Beyond these prominent websites, however, there are a number of social networking sites emerging that are geared specifically towards television.

For example, one popular site, *TVTag*, allows users to "check-in" whenever they start watching a show. Not only does this let someone easily tell their friends about what they're watching in real time, but it also allows them to earn points and rewards.[26]

Second Screen Technologies in Television

Beyond social networks entirely, there are some even more exciting social TV opportunities that have, until recently, gone unexplored. One such opportunity is that of an application that runs on a user's mobile device, determines what they're watching on TV, and then displays content related to what's currently happening on the show. Some promising innovators competing in this area include *IntoNow*, *Audible Magic*, and *Pluk*, just to

---

[26] "tvtag - tag along with the world as you watch TV." *tvtag - tag along with the world as you watch TV*. N.p., n.d. Web. 23 Mar. 2014. <http://tvtag.com/about>.

name a few.[27] At this point, there's no major frontrunner in this space, but it seems likely that this type of technology will break into the mainstream in the near future.

Another type of second screen application that's shaping up to be a hit in the social TV arena is the "gamification" app. For readers unfamiliar with the term, gamification refers to "the process of game-thinking and game mechanics to engage users and solve problems". [28] Gamification has been catching-on rapidly in recent years, and social TV is no exception. In the context of social TV, gamification means that instead of merely watching a TV show, a viewer will be participating in some sort of activity related to the show that has elements of a game.

---

[27] "The Race For The Second Screen: 5 Apps That Are Shaping Social TV."*Co.Create | creativity + culture + commerce*. N.p., n.d. Web. 4 July 2013. <http://www.fastcocreate.com/1679561/the-race-for-the-second-screen-5-apps-that-are-shaping-social-tv>.

[27]"About Audible Magic." *Audible Magic - Digital Fingerprint Content & Media Recognition*. N.p., n.d. Web. 4 July 2013. <http://audiblemagic.com/company.php>.

[27]"pluk." *pluk*®. N.p., n.d. Web. 4 July 2013. <http://www.pluk.com/>.

[28] Zichermann, Gabe, and Christopher Cunningham. *Gamification by design: implementing game mechanics in web and mobile apps*. Sebastopol, Calif.: O'Reilly Media, 2011. Print.

For example, viewers may be invited to predict what's going to happen next. The viewer may then be rewarded for correct prognostications. This type of gaming style is particularly applicable to sports, but there are many other game mechanics that have proven to be usable in nearly every other genre of television. There are countless websites, applications, and services all interested in becoming the leader in this space, but a few noteworthy contenders include *Insticator*, *OneUp Games*, and *PrePlay Sports*.[29] Again, this is just a small fraction of what's out there, and I encourage readers to explore other emerging technologies in this area.[30]

We've discussed what's been popular for social TV over the past few years and what trends have emerged recently. We now turn our attention to the future of social TV and examine what possibilities lie ahead for this growing fusion of social networking and television.

---

[29] "About Us." *OneUp Games | Making Live Sports More Fun!*. N.p., n.d. Web. 13 July 2013. <http://1up.fm/about/>.

[29] "Insticator | Predictions." *Insticator | Predictions*. N.p., n.d. Web. 13 July 2013. <http://www.insticator.com/>.

[29] "PrePlay." *PrePlay*. N.p., n.d. Web. 20 July 2013. <http://www.preplaysports.com/>.

[30] An excellent starting point for further research on this topic can be found at: http://www.lostremote.com/social-tv-companies.

When attempting to decipher what may sit on the horizon in the realm of social TV, we naturally begin by examining current developments. As discussed in previous paragraphs, allowing viewers to share shows they watch and what their thoughts are has been fairly successful. Gamification is another huge area as of late, and the trend doesn't seem to be diminishing. With these two trends in mind, it's likely that any so-called "big winners" in social TV will combine both of these elements in a way that's flawless and easy to use.

Make no mistake, getting both of these elements correct is no small task, and the technological challenges associated with synchronizing the second screen with the TV can be enough to deter even the most experienced software developers. Nevertheless, it seems that anyone who can manage to seamlessly unify these social TV features under an intuitive interface will have an excellent shot at success in this area.

Other innovations we may see emerge in this area include allowing viewers to influence shows in real time, modify the content being shown on TV, or interact with the programs in novel ways. A brief discussion of each of these topics follows.

Users influencing shows via social media has been happening for years. However, the use of social media to influence how a program progresses in *real time* is an area that's ripe for exploration. Immediately, the possibilities for game shows and televised contests come to mind as the obvious target for this type of innovation. That is, rather than asking viewers to call in and vote over the next few hours (or days), the viewers would get a few minutes to vote (during a commercial break, perhaps) and then find out the results almost immediately. This minor tweak allowing for near-instant gratification would most likely be enough to spark the interest of many viewers.[31]

The ability to modify content that users see on their television is not something that's currently possible using traditional television broadcast technologies. That is, one broadcast goes out to many people and it's not feasible to change the content for one viewer without

---

[31] After this section was written, this type of real-time voting mechanism was put into practice in the popular television series, *The Voice*, when they introduced their "Instant Save" segment.

"Thousands tweet to save singers on @NBCTheVoice | Twitter Blogs." *Twitter Blogs*. N.p., n.d. Web. 11 May 2014. <https://blog.twitter.com/2013/thousands-tweet-to-save-singers-on-nbcthevoice>.

changing it for many. Yet, this does not mean that it's impossible, especially when you consider what technologies may emerge in the coming years. Nevertheless, the idea is that the viewer would be able to adjust camera angles being shown on their television set to tailor the viewing experience to their tastes.

For example, instead of seeing a close-up shot of whoever is speaking at the moment, the viewer may decide that it would be more interesting to see a shot of the audience. Or, rather than viewing a wide view of the entire field during a football game, the viewer may want to focus on a shot of the quarterback barking orders right before a snap. In addition, the user may even bring up one angle on their device while watching another on the main screen.[32] The possibilities are both exciting and nearly endless. However, it should be reiterated that this type of innovation is going to require some serious technological support on the backend. Today's broadcast media infrastructure just isn't set up to accommodate

---

[32] Watching an alternative angle on a mobile device is already possible during some sporting events, such as NFL games. However, at present, the viewer cannot change which angle is being shown on the main screen and there are usually a very limited number of options available. In other words, there's plenty of room for improvement and further innovation.

such an experience. Perhaps this may become more realistic as television transitions to a more Internet-based medium in the future.[33]

At present, viewers who wish to interact with shows are typically limited to writing a message or sending in a video to the producers. In the future, it's likely that we will see viewer interaction reach new levels. For example, we could see a game show where the contestant pool is a list of people who signed-up online and are watching from home. The host could then pick a contestant who would play the game live from home via Skype. Even if the chances of a viewer being picked to play were infinitesimally small, the mere fact that a chance exists would likely be enough to keep the interest of many viewers.

This type of social interaction is not limited to game shows, however. Entire programs could be centered on allowing "regular people" to video chat with sports stars, politicians, musicians, and other prominent figures. Again, this may interest viewers and hold their attention more than the average show simply because of

---

[33] There is more on this subject coming up!

the excitement surrounding the idea that they might be the next person to be featured and speak to a celebrity guest.

Social TV is a hot topic right now. Many people believe it's the obvious next step for television, but others think that it may be overhyped and is a fad doomed to fizzle out. Regardless of which side you find yourself on, it can't be denied that there's a significant effort currently being made to unify social media with the television experience. Whether it be the major social networks, niche TV-centric social websites, or even the many second screen apps that have emerged recently, it seems that nearly everyone is trying to cash-in on social TV. Therefore, while it's possible that the phenomenon known as "social TV" may fade away quietly, it seems that there's simply far too much interest in this space for it *not* to continue and be at least mildly successful. In fact, I would expect to see social TV not only continue growing, but to thrive in the years ahead.

Perhaps we may even see some of the more ambitious innovations mentioned previously (such as being able to control camera angles, game shows with contestants at home, etc.) emerge. They won't appear overnight, but over the long haul, I would be more surprised if they *did not* occur. Social TV is in its infancy and is poised to transition into a huge part of the average entertainment experience. Watching this transition occur may end up being just as much fun as the end result. I encourage you to enjoy the ride!

Web-Based Television

Another transition that we're just starting to experience in the realm of TV is the shift from cable, satellite, and over-the-air programming to web-based programming. That is, rather than watching shows via one of these traditional providers, viewers are starting to turn to emerging Internet-based television show providers such as Netflix, Amazon Instant Video, or Hulu. In fact, there's even a small, but growing crusade to sever ties with these traditional providers: the "cord-cutters" movement.

People in this group encourage others to abandon their cable or satellite subscriptions in favor of the cheaper web-based alternatives. At present, online services aren't a perfect replacement for a traditional cable or satellite package, as they typically contain a limited selection of shows that have aired months or years prior to their release on the web. In addition, there's a noticeable lack of live programming such as sports, news, or other televised events on web-based television. These drawbacks have helped limit the "cord-cutting" movement thus far, but this may all start to change very quickly.

One service that's leading the way in the movement towards web-based video content is Netflix. As mentioned, Netflix is a pioneer in the web-based video arena and has become a favorite with many looking to swap their monthly cable bill for something cheaper. More recently, Netflix has further incentivized potential customers to sign-up by offering original content. High-profile releases include *House of Cards*, starring Kevin Spacey, *Orange is the New Black,* and the revival of cult-hit *Arrested Development*. All three of these releases garnered quite a bit of attention before and after their release. In addition, fan reaction has been mostly positive.[34] This type of buzz has encouraged other online

---

[34] "Arrested Development." *IMDb*. IMDb.com, n.d. Web. 23 Mar. 2014. <http://www.imdb.com/title/tt0367279/?ref_=nv_sr_1>.

video services to create and release their own original content.

Amazon and Hulu have also started to roll out their own original content and it appears that the general public's consensus has been mostly enthusiastic. In other words, the transition of web-based video services from syndicated content providers to full-fledged, online television channels competing directly with traditional television is happening right before our eyes.

This transition may not seem significant at first glance. However, if current trends are taken into consideration and we consider where this may lead, it becomes a real possibility that web-based programming will become a true competitor with traditional television providers. In fact, given the right circumstances, it's not outside the realm of possibility to imagine a day where web-based programming becomes the primary way people watch TV.

Web-based services offer features that just can't

---

[34] "House of Cards." *IMDb*. IMDb.com, n.d. Web. 23 Mar. 2014. <http://www.imdb.com/title/tt1856010/?ref_=nv_sr_1>.

[34] "Orange Is the New Black." *IMDb*. IMDb.com, n.d. Web. 23 Mar. 2014. <http://www.imdb.com/title/tt2372162/>.

be matched by traditional television, such as being able to watch your show from a multitude of devices and from nearly any physical location that has Internet access. You can't plug a satellite dish into your tablet or attach the TV in your hotel room to your cable subscription while out of town -- it just isn't possible.[35] In addition, web-based television providers offer viewers the ability to browse a library of shows, pick and choose *which* shows they want to watch, and decide *when* they watch them. Again, these types of features simply aren't supported by traditional TV content delivery infrastructures. For these reasons, it seems rather clear that web-based television is practically destined to supplant traditional providers in the years ahead.

Of course, Internet-based television won't beat traditional TV only by having superior features. As the saying goes, "content is king," and I suspect most people won't trade their cable or satellite subscriptions for a web-based alternative with sub-par content. However, it

---

[35] Technically, you *can* access your home cable or satellite television content using a special device such as a Slingbox. However, these devices allow access to this content via the Internet. Thus, it still demonstrates the limitations of traditional cable and satellite, while highlighting the flexibility of Internet-based services.

appears that this isn't a decision that many will have to face, as both the quality and quantity of online television offerings has been steadily increasing, with no signs of decline on the horizon.

New content providers are emerging at a breakneck pace accompanied by new devices and gadgets that aim to improve and transform the traditional television experience. There are countless such services and devices, but there are a few notable examples that stand out of the crowd as true frontrunners in this space.

The obvious choices in the service provider arena are Netflix, Hulu, Amazon Video, and YouTube. These are household names, and for good reason -- they offer exceptional content at a reasonable price. There are, however, some lesser-known services out there that may be able to fill some of the gaps left by these mainstream frontrunners.

One such service is *Aereo*, which is a newcomer compared to the other more well-known video services. However, Aereo has managed to gain an impressive following in a little over a year, while ruffling quite a few feathers along the way. Somewhat uniquely, rather than offering users a selection of videos to choose from on-demand, Aereo provides customers with live streams of local channels that can be accessed from nearly any Internet-connected device.

Typical offerings of Aereo include Fox, CBS, NBC, CW, and perhaps a few public access channels. While it's true that these channels are broadcast and can be viewed for free by anyone in the area with an antenna that receives the channel, many people simply don't have good reception in their homes due to mountains, valleys, tall buildings, or other large objects standing in the way. For many years, people who found themselves unable to receive local channels over the air were either forced to pay for a cable subscription or abandon hope of watching these channels. This type of person would greatly benefit from Aereo and its relatively cheap monthly fee.[36] Similarly, anyone who frequently travels or otherwise watches television from a laptop, tablet, or mobile device may find Aereo particularly appealing.

As mentioned, Aereo has faced some pretty tough

---

[36] At the time of writing, Aereo's basic monthly fee is $8 per month. "Frequently Asked Questions" *Aereo*. N.p., n.d. Web. 20 July 2013. <https://aereo.com/faqs>.

opposition to their novel service by broadcast networks and the cable companies alike. While Aereo did score a pretty big legal win recently when a judge ruled that Aereo was not guilty of copyright infringement, there are surely many more legal challenges ahead for this video streaming pioneer.[37] If the company can successfully fend off future litigation, it may end up becoming a favorite in the online video service community.

Another type of web video service that's been gaining some momentum gives users their own channel, from which they can stream whatever video feed they wish. It can be a webcam stream of someone just talking to viewers, a live video feed of someone playing a video game, or even a stream of TV shows or movies that the user is currently watching. Visitors can then browse through these channels and enjoy a wide range of non-traditional television programs free of charge.

One of the most well-known services in this space is *JustinTV*. Other popular options include *Veetle*,

---

[37] While this book was in its final stages of editing, Aereo's legal battle was taken to the next level when the U.S. Supreme Court decided to hear the case against the company. A win for Aereo here could solidify the company as a major player in the next generation of media entertainment. A loss would likely result in the end of the company.

Ustream, and *Freedocast.* The legality of these services will obviously vary depending on what the user is streaming and where the user is located geographically. That is, streaming copyrighted content may constitute copyright infringement and could land all involved parties in legal trouble. Nevertheless, the legal streams provided by these services offer viewers a fairly nice entertainment alternative to traditional TV that doesn't cost them a dime.

A third type of video service that has become rather popular with certain groups of people allows users to view nearly any TV show or movie free of charge. The content is typically listed on sites dedicated to "media pirates" in the form of web links. Obtaining the movies or television shows is as simple as clicking the link. While this is certainly appealing from a television viewer's perspective, these services are nonetheless illegal in most developed nations, as they aren't sanctioned by the legal owners and do not generate any revenue for the content creators.

The legal and ethical issues presented by media piracy and copyright infringement are abundant and complex enough to fill many books dedicated entirely to those topics, so we won't go into much discussion here. It's worth noting, however, that if content creators aren't receiving any financial compensation for their works, then they won't be able to continue creating them.

These types of services won't be listed here, and I would encourage all readers to support the creators of the content they enjoy! Still, it's important to mention that these services exist, are used by quite a few people, and are an alternative that people use to obtain and view their favorite television shows. This type of consumer behavior will undoubtedly continue to impact how television services evolve.

The Role of Mobile in Next-Gen TV

The ability to watch TV online is something that many people are excited about. Nevertheless, being able to choose *how* and *where* a show is watched is sometimes just as important as the show itself. That is, even if people can access all of their favorite shows at the click of a button, the experience still leaves much to be desired if they must sit at a desk in an office chair to use their desktop computer. Thankfully, there's been an enormous amount of progress in this area over the last few years. Today, Internet-based television can be watched from nearly every location including the old standby: on the living room "big screen TV" from the comfort of the couch.

There are not *quite* as many devices that consumers can choose from to access web TV, but new devices are emerging regularly and the selection is fairly diverse. From mobile devices such as smartphones and tablets to standalone units designed to attach to a television set, and everything in between; there are enough options that nearly everyone can find something that fits their needs. Each of these options offer slightly different experiences and come with their own unique pros and cons.

Smartphones and tablets comprise the newest category of devices capable of delivering TV shows to viewers. A decade ago, neither of these devices even existed among the general public. Today, consumers can choose from thousands of portable gadgets, from which they can enjoy countless entertainment options.

The list of hardware manufacturers is quite lengthy, but most of these devices run one of four operating systems: Android, BlackBerry OS, iOS, or

Windows. These devices can playback nearly all of the same content as a desktop computer, but with the added benefit of portability. Using one of these gadgets, a person can watch "television" from just about anywhere. Of course, if the user wants to access a web-based video service, they must also be in an area that has Internet connectivity. There's also typically no easy way to connect to a physical media source, such as a DVD or Blu-ray disc player, to these devices. Furthermore, watching video drains a battery rather quickly which means a user will have connect to a power source after a short amount of time. This effectively diminishes the main advantage of these portable devices.

Still, it can be argued that some portability is at least better than none at all. This seems to be a viewpoint shared by most consumers, as sales of portable devices have been increasing over the last decade, while sales of desktop computers have been on the decline.[38] All in all, the future of accessing television and other videos on smartphones and tablets looks very promising, especially

---

[38] "Global Tablet Shipments to Overtake PCs by 2015, IDC Says." *Bloomberg - Business, Financial & Economic News, Stock Quotes*. N.p., n.d. Web. 7 Sept. 2013. <http://www.bloomberg.com/news/2013-05-28/pc-market-to-decline-7-8-in-2013-as-mobile-devices-gain.html>.

when considering short battery life will become less pronounced as related technologies improve.

A second type of device that consumers are gravitating towards allows users to watch their web-based television content on a TV set. Believe it or not, accessing the web from a television is something that has been occurring for nearly 20 years. The first commercial attempt appeared in 1995 with a device known as "WebTV".[39]

Perhaps WebTV was a bit ahead of its time, as it never fully caught on. Nevertheless, the idea of connecting to the Web from a television set had been put on the map. Two decades later, this idea has been revived in a big way and is gaining traction once again as the popularity of video-on-demand services continue to increase.

Today, the average consumer who wishes to access web-based content from a television can choose from dozens of devices. Some of the most popular options include Apple TV, Boxee Box, Google TV, and

---

[39] "Steve Perlman and WebTV (A)." *Harvard Business Review Magazine, Articles, Blogs, Case Studies, Books - Harvard Business Review*. N.p., n.d. Web. 7 Sept. 2013. <http://hbr.org/product/steve-perlman-and-webtv-a/an/899270-PDF-ENG>.

Roku.[40] We won't cover each of these individually, as they all have their own benefits, drawbacks, and quirks. Nevertheless, the main purpose of each is to allow a user to enjoy web-based, as well as locally-stored video content on their televisions. This means that the average consumer can relax in their living room and have thousands of entertainment options available at the touch of a button, on-demand, and in high-definition. Combine that with the fact that these units typically sell for under $100, and it's not hard to see why they've become extraordinarily appealing for cord cutters or anyone who wishes to trim their monthly entertainment bill without sacrificing too much in terms of the quality or quantity of entertainment options.

A third category of device that provides users with the ability to watch television is the video game console. The idea of a console being able to provide top-notch TV and movie entertainment may come as a surprise to some people, particularly those who grew up in an era where the pinnacle of game graphics were the likes of *Donkey Kong Country* and *Sonic the Hedgehog*.

---

[40] NOTE: Google TV software runs on a variety of devices manufactured by third parties.

Nevertheless, consoles have come a long way in the last twenty years. Not only have games improved, but the types of entertainment a video game console can provide increased somewhat dramatically, as well.[41]

For quite a few years now, console makers have been adding more functionality and entertainment options than what was traditional. The transition from such hardware being single-purpose devices to multi-purpose machines started to hit its stride in the year 2000, when Sony introduced the extraordinarily well-received DVD playback feature with the PlayStation 2. A decade later, the next phase of this transition is well underway, with most consoles readily able to connect to video-on-demand services. This allows users to play games and watch videos on their televisions, all with the same device.

It should be noted that in the latest generation of gaming consoles, two devices -- the Wii U and Xbox One -- allow users to view live television as well. This means that gamers can access video-on-demand, video games, and live television all from one entertainment device. To

---

[41] More on in this subject coming up!

be sure, these devices are both still fairly young and this technology will need some time to fully mature. Nevertheless, if this tech is developed correctly, this move could turn out to be a gigantic step forward in the transition from *video game* consoles to *entertainment* consoles.

In addition to the conventional video game device, there are also "handheld" or "portable" gaming gadgets. Most readers have probably seen at least a few of these devices, which started gaining popularity and momentum in the early 1990s with the release of the original Nintendo Gameboy in 1989.[42] There have been many iterations between that time and today, but the two most popular handheld devices available today are Nintendo's 3DS and Sony's PS Vita. Both of these devices provide far more functionality than just gaming and offer video-on-demand from a wide variety of sources.

A chapter of this book is devoted entirely to video gaming, so we'll postpone most discussions related to these consoles until then. For the purposes of this section, however, these consoles are noteworthy

---

[42] "Nintendo - Corporate Information | Company History." *Nintendo*. N.p., n.d. Web. 7 Sept. 2013. <http://www.nintendo.com/corp/history.jsp>.

because they give consumers yet another option of when, where, and how they can watch their favorite content.

With so many ways for the average person to access web-based video services, one starts to wonder why anyone would still be paying for a subscription television service.

The answer to this question is that there are still many limitations to the types of content offered by video-on-demand services. Nevertheless, most services offer far more hours of entertainment than anyone could ever need, at a fraction of the price of a traditional cable or satellite subscription. With this in mind, the future of these cable and satellite providers doesn't look promising. While it's true that they offer programming that can't be matched, such as exclusive channels and content, these offerings may not be enough to hold onto customers, as more people seem to be exploring cheaper web-based alternatives.

The exact data related to how many people are opting out of cable is still somewhat fuzzy. In 2011, *Huffington Post* ran an article with the headline, "Cable and Satellite TV lose Record Number of Subscribers."[43] In August of the following year, it was reported that Comcast, one of the largest cable providers in the U.S., lost nearly 400,000 cable subscribers between 2011 and 2012.[44] In March of 2013, a different report indicated that the number of cable subscribers *increased* overall in 2012.[45] An article published in Forbes concluded that although the number of people abandoning cable -- or who don't subscribe in the first place – isn't meaningful yet, it "does seem to be the beginning of a 'macro trend' that will ultimately change the business radically."[46]

There are countless websites, articles, message

---

[43] "Cable And Satellite TV Lose Record Number Of Subscribers." *Breaking News and Opinion on The Huffington Post*. N.p., n.d. Web. 10 Aug. 2013. <http://www.huffingtonpost.com/2011/08/10/cable-tv-satellite-loses-subscribers_n_923034.html>.

[44] "Comcast Loses Nearly 400,000 Subscribers In Last Year." *Business Insider*. N.p., n.d. Web. 10 Aug. 2013. <http://www.businessinsider.com/comcast-loses-nearly-400000-subscribers-in-last-year-2012-8>.

[45] "Cord Cutting: Pay TV Subscriber Growth Up for 2012, Allaying Concerns." *The Latest Entertainment & Hollywood News - The Hollywood Reporter*. N.p., n.d. Web. 10 Aug. 2013. <http://www.hollywoodreporter.com/news/cord-cutting-pay-tv-subscriber-426341>.

[46] "Are Cable TV Carriers Seeing Meaningful Subscriber Degradation Due To Young People Not Signing Up? - Forbes."*Information for the World's Business Leaders - Forbes.com*. N.p., n.d. Web. 10 Aug. 2013. <http://www.forbes.com/sites/quora/2012/06/20/are-cable-tv-carriers-starting-to-see-meaningful-subscriber-degradation-due-to-more-young-people-not-signing-up-for-cable-or-satellite-tv-in-their-homes/>.

boards, and discussion threads dealing with the debate over whether or not the cable industry is heading towards tough times as a result of so-called "cord cutters." The truth of the matter is the data just isn't conclusive enough yet to determine whether these reports of declining cable subscribers are temporary setbacks or signs of a cultural shift. While the latter of these two possibilities seems more likely, the only surefire way to find out is to sit back and observe the future of TV unfold. Even if only half of the innovations discussed in this section reach their full potential, watching how this industry progresses may turn out to be just as entertaining as television shows themselves.

Discussion Questions:

1) Have you participated in the social media aspect of a television show you've watched?

If so, describe the type of social interaction that you engaged in. What was it about this type of show and social media component that compelled you to participate in the "social TV" experience? Explain.

If not, do you plan on participating in the future? Is there any particular type of "social TV" experience that would compel you to engage in the practice? Explain.

2) The author suggests that, in the future, viewers may be able to pick and choose which camera views are shown on their television. Do you agree that this is a likely next step for the industry? Why or why not?

3) Do you agree with the author's assessment that more and more television will be broadcast over the Internet? Why or why not?

4) Do you want to see more television broadcast over the Internet? Why or why not? Explain.

5) What impact do you see the cord-cutting movement having on the industry? If you are a "cord-cutter" describe your experience and

what motivated you to join this group. If you aren't a cord-cutter, what conditions would cause you to join this movement? Explain.

6) If it was legal to download television shows and movies for free, do you think this content could still earn enough money to be financially viable? If yes, describe how you envision this type of content earning revenue. If not, how do you envision programs earning revenue in the future? (For example, selling ads, paying per view, both, etc.)

7) If you were designing a new video-streaming device (a competitor of Roku, Apple TV, and Google TV), what futuristic features would you like to see incorporated in the final product?

8) Do you think the trend of subscribers abandoning their cable subscriptions will continue? Why or why not? Explain.

Chapter 5: Music

Piracy and Turmoil

When determining how technological innovation may impact the music industry in the years ahead, it's important to recognize just how much technology has affected the creation, production, and distribution of music over the past decade. File-sharing flipped music distribution on its head, digitally-synthesized music reached new heights, and studio-quality recording software is now available to anyone with a computer and a microphone. Nearly everyone involved with music has felt the impact of technology on the industry in recent years.

Perhaps the largest change in music production and distribution in recent memory was the introduction of file sharing to the general public. This topic was covered fairly extensively earlier in this book, and there are quite a few more interesting areas of technological innovation in the realm of music, so we won't explore the topic much in this section. Nevertheless, it's worth briefly mentioning the type of impact this has had, since it's affected nearly every aspect of the industry.

When file sharing started gaining popularity with

the general public around the turn of the millennium, the music industry wasn't quite sure how to best respond. Over a decade has passed, and the industry *still* doesn't have a surefire way of preventing people from obtaining music online without charge.

To be sure, there have been many lessons learned in the past decade. For instance, the once-popular tactic of suing copyright-infringers has been retired after years of yielding minimal results.[47] Nevertheless, large-scale file sharing still occurs, sales are declining, and music industry executives have been left scrambling to put together a new plan.[48]

A decade later, it's fairly easy to see that the tech which lead to file-sharing and music piracy was (arguably) the technological innovation that had the largest impact on the music industry in that time. With no clear solution

---

[47] Young, Jeffrey R.. "Music Industry Will Stop Suing Groups of Students - Wired Campus - The Chronicle of Higher Education." *Home - The Chronicle of Higher Education*. N.p., n.d. Web. 7 Sept. 2013. <http://chronicle.com/blogs/wiredcampus/music-industry-will-stop-suing-groups-of-students/4448>.

[48]"RIAA - Anti Piracy - September 07, 2013." *RIAA*. N.p., n.d. Web. 7 Sept. 2013. <http://www.riaa.com/faq.php>;"Three Strikes For File-Sharing Fails to Halt Music Sales Decline." *TorrentFreak | Breaking File-sharing, Copyright and Privacy News* . N.p., n.d. Web. 7 Sept. 2013. <http://torrentfreak.com/three-strikes-for-file-sharing-fails-to-halt-music-sales-decline-130601/>.

in sight, it's likely this impact will carry on, with permanent effects on the shape of the industry.

While the technology that enabled file sharing may have had the largest impact on the music industry in recent memory, it's far from the only innovation that has helped shape where the industry is today and where it's headed tomorrow. There are plenty of other developments that have been introduced into the industry over the past few years that are both more exciting and interesting to discuss. One such innovation is streaming music.

Streaming Music

The idea of streaming music and other audio feeds across the Internet has grown in popularity since the mid-1990s, when the concept was introduced to the general public.[49] Since then, the technology behind streaming audio has been used to live-broadcast countless concerts, speeches, and events across the globe. In addition, pioneering companies like *RealNetworks* used this technology to recreate the audio

---

[49] "Musical Events." *Kevin Savetz*. N.p., n.d. Web. 7 Sept. 2013. <http://www.savetz.com/mbone/ch6_4.html>.

broadcasting experience traditionally found on terrestrial radio stations. Once this groundwork was laid, Internet-based radio exploded in popularity. In fact, web radio is reportedly on the cusp of surpassing terrestrial radio; which may in fact have already happened by the time this book goes to print.[50]

In addition to recreating the traditional radio experience, online music streaming has paved the way for novel *radio-like* experiences that weren't possible a few decades ago. Services like *Pandora*, *Rhapsody*, *Slacker*, and *iHeartRadio* all provide the listener with a stream of music akin to the traditional radio station format, but with added abilities for the user to interact with and customize the station. If the user doesn't want to listen to the song currently being played, they can skip it or indicate that they don't want to hear it again. The more the user interacts with the service, the more it learns their likes and dislikes. This greatly improves the user's experience and provides a major advantage over

---

[50] Gilbert, Jason. "Internet Radio Is About To Surpass FM/AM, But Have Spotify And Pandora Really Changed Anything?." *The Huffington Post*. TheHuffingtonPost.com, 9 Nov. 2012. Web. 23 Mar. 2014. <http://www.huffingtonpost.com/2012/11/09/internet-radio-is-about-t_n_2101311.html>.

traditional radio, which simply doesn't have the infrastructure to match this type of feature.

With all of this in mind, it may be tempting to conclude that this evolution of Internet radio will undoubtedly be the most popular type of audio streaming service in the coming years. However, a very different type of streaming solution has emerged which has gained quite a few supporters in a short period of time: selectively streaming only the songs, artists, or albums chosen by the user. This type of service started gaining momentum when the *Spotify* music service debuted in 2008, and has only continued to thrive as more audio streaming services have added similar capabilities.

The main selling point of this type of solution is that users can pick and choose exactly what they want to hear, while excluding everything else. In many ways, these services mimic the user experience that accompanies downloading and managing music on a hard drive stored at home, while reducing the hassle of managing a music collection.[51] What's more, most of

---

[51] Anyone who spent any amount of time transferring music off of old CD's

these services either have a free trial period or even a free option. All of these features add up to a streaming service that is *very* competitive with the more common Internet-based radio options. However, these services have their own drawbacks.

For example, most of these services only work if the user is connected to the Internet. If the Internet goes belly up or a user is in a location without Internet access, there won't be music until a connection is re-established. Furthermore, most of these services don't allow free streaming to mobile devices. This means that if a user wants to listen to music on the go, they may have to pay for a premium plan. Nevertheless, the popularity of these services demonstrates that selectively streaming music may have the potential to become the standard delivery method this decade. Of course, only time will tell how popular opinion will evolve.

One possible area of innovation related to this area deals with the idea of blending of local and social content into custom-tailored online radio stations.

Currently, users can either listen to a live stream

---

will likely know all about the challenges of sorting through, tagging, and managing hundreds of songs completely void of any metadata.

of a local radio station online or an Internet-based station which plays songs based on user-feedback. Local stations feature live radio personalities discussing the news, weather, and other entertaining topics.[52] Hearing about local community news and events is a major focal point for most terrestrial radio stations and almost certainly plays a role in keeping them alive in this digital era. However, being able to custom-tailor songs to a listener's liking clearly has advantages over whatever the station happens to play at a particular moment.

Thus, an opportunity exists for a music service to offer the ability to seamlessly blend live local or nationally syndicated radio programs with customized music playlists. Listeners will hear radio personalities discuss the daily news and events, and then a different song will be played for each user depending on their personal preferences. For users that want to hear the news *and also* want to pick which artists, albums, or songs to listen to, this type of functionality would seemingly offer the best of both worlds.

However, there are clearly some technical

---

[52] Sometimes!

limitations to consider. For example, since each user will be listening to a different song, it will be nearly impossible for each user to hear the radio personality in real-time. The obvious solution to this would be to have some listeners hear the disc jockey's banter on a delay. This does present some new issues, though, such as how to synchronize the listeners for call-in contests or other listener-oriented segments. Nevertheless, these issues aren't insurmountable, and even if some of these limitations couldn't be fully overcome, the end-result would still be a much better blend of features than anything that exists today.

A limiting factor to consider, which may slow progress in this area, is that this type of setup wouldn't be possible on local terrestrial radio. Similar to the discussion about television broadcasting earlier in this book, radio stations aren't able to broadcast different content to each user due to the way that radio waves behave.[53] This means that this type of radio station could only exist online. Nonetheless, as we transition from analog radio to its digital counterpart, this particular

---

[53] This limitation is essentially identical to what is experienced in television broadcasting, which was discussed earlier in this book.

limitation may become less pronounced.

Even with these limitations in mind, there aren't any major *technological* hurdles standing in the way of this type of radio station being created. In fact, the only real issue that may prevent such a service is the decision of whether or not the broadcasting rights for such a station could be secured. Although, this new type of "hybrid radio" could still include all of the same commercial slots and bring in similar advertising revenue as traditional radio, so there are no obvious roadblocks in this area.

In fact, this type of station may actually bring back listeners who have started to abandon local stations in favor of Internet radio. After all, this type of service would be the "best of both worlds," providing the listener with music they want to hear with just a touch of local flavor. There are, of course, many challenges and other potential roadblocks in the way of making this type of service a real success, but for an innovation that has yet to be introduced, there seem to be surprisingly few technical challenges.

The Next-Generation Physical Audio Format

With all of this talk about Internet-based radio, selective streaming options, and music downloading, it's easy to forget that there's yet another way for the average consumer to obtain their favorite music: purchasing a physical CD.

By today's technological standards, the compact disc is a dinosaur. First released in 1982, the CD is now over 30 years old, with design aspects that can be traced back to at least a decade earlier than that.[54] Nevertheless, the CD was superior in many ways to the cassette tape, which the CD surpassed in the early 1990s to become the best-selling physical audio format of the era.[55] The popularity of the CD continued to grow throughout the 1990s until the end of the decade, when

---

[54] Sawers, Paul. "The CD is 30 Years Old Today." *The Next Web - International technology news, business & culture*. N.p., n.d. Web. 17 Aug. 2013. <http://thenextweb.com/media/2012/10/01/the-first-commerically-available-cd-album-player-released-30-years-ago-today/>.

[54] "The emergence of the compact disc."*IEEE Xplore - Home*. N.p., n.d. Web. 17 Aug. 2013. <http://ieeexplore.ieee.org/xpl/login.jsp?reload=true&tp=&arnumber=5394021>.

[55] Lawrence, Ross, (2010). "Can the music industry adapt to the digital future?" *Econ Focus*, issue 3Q, p. 26-27, http://EconPapers.repec.org/RePEc:fip:fedrrf:y:2010:i:3q:p:26-27:n:v.14no.3.

file sharing hit the mainstream. Since the turn of the millennium, CD sales have consistently declined year over year, leaving many to wonder if the CD is finally nearing retirement.

If the CD is indeed reaching the end of its shelf life, it raises the possibility that the CD may be the last physical music storage format of its kind. That is, CD sales have been losing market share to digital copies for over a decade now.[56] This sends a message to the industry that consumers prefer the convenience of a digital download. The question that's undoubtedly left on the minds of music industry insiders is whether this indicates consumers prefer obtaining their music digitally as opposed to *any* physical medium. Or however, that consumers *only* prefer digital copies over the outdated compact disc.

This may not seem like a matter of much importance, but misinterpreting the public's true opinion on this issue could result in a fairly catastrophic misstep for an industry at a particularly fragile point in its history.

---

[56] "Digital Music News - The History of Recording Industry Sales, 1973-2010...."*Digital Music News - Home.* N.p., n.d. Web. 2 Sept. 2013. <http://www.digitalmusicnews.com/stpries/021711disruption>.

That is, if a small fortune is invested in the creation of the next audio standard, but it doesn't catch on with the public, it could turn into a financial disaster. On the other hand, if consumers desire a new, smaller, and more durable physical medium, but the industry continues to produce CDs, it may drive more people to digital downloads, streaming services, or away from purchasing music entirely. Determining whether or not to release a new physical music format is a critical issue that may very well decide the fate of the music industry.

Even if the public ultimately does desire a new physical format, there's an even greater question that must be answered: what will this format be? At present, the only real contender is the USB flash-drive. Most readers are probably familiar with these tiny devices that also go by the name "thumb drive," "USB stick," or "jump drive." These devices are more compact, more durable, and offer far greater capacity than a CD. Furthermore, because of their increased space, these devices could easily contain higher quality audio than what's found on CDs today.

The only real downside to the potential of this format is that it has already been tried on a small scale

107

since as early as 2008, but hasn't received a particularly warm welcome from the public.[57] Whether or not public opinion on this issue changes in the years ahead still remains to be seen. Nevertheless, it does seem entirely possible that the CD may turn out to be the last widely-used music format that doesn't reside entirely in the digital realm.

Technology's Impact on Musicians

In addition to changes in distribution, we're also in the middle of a shift in how artists are "discovered" or "make it big" in the music industry. A few decades ago, the standard formula was to perform at every venue possible and hope to catch the attention of a record label insider. Once an artist was signed to a label, they could count on a tour, merchandise, and a record deal, of course. This isn't to say that it was a free ride or particularly easy to become a "star." However, being signed to a record label was undoubtedly the most common way to the top.

Today, with the rise of Internet-based music

---

[57] "Replacement for CD nowhere to be found." *APC - Australian tech news, how to and computer help.*. N.p., n.d. Web. 24 Aug. 2013. <http://apcmag.com/replacement_for_cd_nowhere_to_be_found.htm>.

distribution vehicles like YouTube, MySpace, ReverbNation, and Last.fm, the role of the record label is changing. In fact, as was mentioned in the introduction to this book, some would even argue that the role of the record label is *disappearing*. That is, with all of these new innovative online services, an artist can accomplish all of the same tasks that used to be reserved for industry insiders without ever signing a contract.

Artists like *Macklemore and Ryan Lewis* seem to reinforce this idea, as they work their way to the top of charts without the support of a label. It's worth noting, however, that even though unsigned artists do crop-up on the music charts every so often, it's still considered somewhat of a rarity for such a feat to occur; they are the exception to the rule.

When considering this, the question of what the future will bring for record labels comes to mind. From a historical standpoint, record labels seem to be a crucial piece of the puzzle that is the music industry. Artists rely on labels for financial support and their vast network of connections to propel them to stardom. This is the way it's been for years, and it'll likely remain this way in the near future. However, when contemplating what lies ahead for labels in ten, twenty, or even thirty years down the road, their role in the industry, and even their existence, becomes less certain.

The quality of home audio recording equipment is constantly coming closer to what has historically only been found in professional-grade recording studios. Online distribution is the fastest, most cost-efficient, and easiest way to release music to the entire connected world. Social networking allows musicians to advertise their art to thousands of people at the push of a button and connect directly to those who can help them take the next step in their career.

Each year, as more of the tools once only available to industry insiders are brought to the fingertips of the general public via the Internet, it starts to become clearer that the record labels are almost certainly destined to become irrelevant to an artist's success or failure in the music industry. After all, if an artist can accomplish the same tasks without the label, then the label is no longer an asset, but rather another obstacle between the artist and success. Make no mistake, labels are entwined with the music industry and aren't going away anytime soon, but as time goes on, it's all but certain that we'll see more independent artists rising to the top without the aid or support of record labels.

Music Production

Aside from music distribution, consumption, and industry politics, there is one other major area that has recently been affected enormously by technological innovation: music creation and production.

As was discussed earlier, the rise of the Internet and Internet-based music distribution has allowed artists to push their music out to more fans easier and more quickly than what was previously possible. This has allowed more musicians to become accessible to potential fans than ever before. This has created a saturation of the music market unlike anything that has ever existed. In turn, this has resulted in artists innovating in new and exciting ways to stand out and gain recognition. Artists are combining genres that don't intuitively blend together to produce some unique and interesting music.[58] In addition, through the use of

---

[58] Two fairly well-known examples that come to mind: 1) Rapper *Nelly* crossing over into country music on multiple occasions in the past decade. 2) Folk-Rock band *The Avett Brothers* blending old-time country tunes with a more modern rock feel.

111

technology, artists are creating new sounds and even entire genres that were not possible a decade ago.

Technology-based music, labeled "electronic," "dub-step," or "EDM" (short for electronic dance music), is on the rise and captivating younger audiences.[59] Interestingly enough, it appears that this new genre may be transforming the idea of what constitutes the quintessential rock star. Whereas the slightly older generation may still think of someone head-banging their way through a solo on their electric guitar, to the younger crowd, the term "rock star" could now just as likely refer to a person up on stage standing behind a laptop or two, tapping buttons and swaying to the beat. Regardless of your musical tastes, it's nonetheless fascinating that through the use of technology, an artist could garner thousands of fans, put on shows around the world, and achieve many of the highest honors in the music industry -- all without necessarily being able to sing, play an

---

[59] "Electronic Dance Music's Rising Popularity...." *NAMM.org*. N.p., n.d. Web. 23 Mar. 2014. <http://www.namm.org/news/press-releases/electronic-dance-music%E2%80%99s-rising-popularity-fuels>.

[59] "The Rise of EDM." *Music Business Journal Berklee College of Music RSS*. N.p., n.d. Web. 23 Mar. 2014. <http://www.thembj.org/2012/10/the-rise-of-edm/>.

instrument, or write lyrics.[60]

For readers who may not be fully convinced that this type of scenario is possible in practice, there are already numerous real-world examples of prominent artists making it big through the use of technology in the electronic music industry.

For example, the artist *Skrillex* began his rise to stardom in 2010 after the breakup of his former non-electronic-based band *From First to Last* in 2007.

Some readers may be vaguely familiar with Skrillex as the person who inspired a generation to shave half their hair and grow the other half down to their shoulders; a style appropriately referred to as the "Skrillex haircut." Nevertheless, this artist has managed to gain quite an impressive following for his creativity in music over the last few years, and not just for his unique appearance.

Skrillex has regularly sold out venues around the world. These shows feature him on stage behind a myriad of electronic equipment twisting knobs, flipping switches and swaying with the crowd. After observing a few

---

[60] Of course, this will depend on whether you consider a laptop loaded with music software an "instrument."

recordings of these shows, it appears that he doesn't sing, rap, or even talk to the audience very often. Throughout his shows, he seems to prefer to let the electronic rhythms and sounds that he's created do the talking for him. This method has worked out well for him so far, as he's managed to score a total of six Grammys and numerous other industry awards in just under three years.[61]

It's worth pointing out that Skrillex is far from the first artist to make it big in the electronic genre and he most certainly won't be the last. There are countless other musicians who could just as easily be used to demonstrate that electronic-based music is on the rise and seemingly destined to redefine how the coming generations think of terms like "concerts," "musicians," "rock stars," and even "music" in general. Each year, it seems more likely that a genre of music that was once relegated to the fringes of the industry is well on its way to becoming the defining genre in the next generation of

[61] "Skrillex on Grammy Wins: 'I Didn't Expect to Be Here Again' | Music News | Rolling Stone." *Rolling Stone | Music News, Politics, Reviews, Photos, Videos, Interviews and More.* N.p., n.d. Web. 2 Sept. 2013. <http://www.rollingstone.com/music/news/q-a-skrillex-on-grammy-wins-i-didnt-expect-to-be-here-again-20130210>.

music.

Since EDM is arguably the genre most influenced by technology, it makes a lot of sense that we'll likely see the most innovation in this genre in the years ahead; at least from a technological standpoint. Since the digital revolution began hitting its stride in the mid-1990s, there's been a steady rise in digitally produced electronic music. Of course, this current brand of electronic music is a far cry from what was created a decade ago, and it's extraordinarily likely that the music being produced and labeled "electronic" a decade from now will only vaguely resemble the music of today. Such is the nature of music in general. However, when it comes to electronic music, the audio production techniques and the resulting sounds change almost as rapidly as the technology on which it is based. This has allowed an enormous amount of innovation in the electronic genre in a very short period of time. What's more, there are some extraordinary technologies just on the horizon that promise to make electronic music one of the most innovative and exciting genres in the decade ahead.

One interesting innovation that's being introduced to the public is an electronic instrument that's a cross between a guitar and a synthesizer. Readers who lived through the 1980s may recall images of the "keytar," which was a popular instrument during that decade. However, although this new generation of digital guitar may resemble the keytar in some respects, they are typically much closer to an electric guitar in terms of appearance, and even closer to being a new instrument altogether in terms of sound.

The so-called "Kitara," manufactured by a small startup called Misa Digital Instruments, was one of the pioneers of this new musical frontier. This instrument started gaining popularity in 2012 when the guitarist of the rock band *Muse* began playing the Kitara during live shows, including a television appearance on *Saturday Night Live*.[62] The Kitara resembles an electric guitar with a few notable differences. First, there is no headstock. The neck of the guitar simply starts at the first fret. Second, there are no strings. Each fret contains six small rectangular buttons that mimic the position a guitarist would typically press the strings. Finally, a touchscreen

[62] "Instrument Muse Played On Saturday Night Live – Status Graphite Bass With Kitara?." *Home Brew Audio*. N.p., n.d. Web. 23 Mar. 2014. <http://www.homebrewaudio.com/instrument-muse-played-on-saturday-night-live-status-graphite-bass-with-kitara/>.

takes the place of audio pickups. By pressing the buttons along the fret board and controlling the sound effects via the screen, the Kitara is capable of producing some interesting and almost entirely unique music. In fact, regardless of musical preference, there is something truly exciting about seeing the capabilities of the Kitara brought to life, and I would encourage all readers to take a brief moment and search online to see this instrument in action.[63]

Sadly, in August of 2013, this particular model was discontinued before it could gain true mainstream support. Nevertheless, it seems to have shed some light on where digital music creation may be headed in the next few years. There are numerous other instruments still on the market that are variations of the Kitara, including Misa's official successor to the Kitara, the Misa *Tri-Bass*. As of yet, none of these models have managed to gain the same level of popularity achieved by the *Kitara,* but endless possibilities presented by this type of instrument leave little doubt that this new generation of

---

[63] The song "Madness," by Muse, was typically played on the Kitara. An interested reader should be able locate a live performance of this song on YouTube or another video-hosting website.

digital guitars have a bright future.

Another brand new instrument that seems destined to make waves in the electronic music scene, and music in general, is the "musical glove." Developed by graduate students at MIT and currently being popularized by English musician Imogen Heap, this device allows for completely unique audio production and performance experiences. That is, not only can these gloves create novel sounds, but due to the way that the user interacts with the device, a work of performance art is created each and every time these gloves are worn to play a song.

The gloves contain sensors that detect movement of the arms, elbows, and hands. In combination with a microphone that captures the vocals, the glove wearer can move their hands and arms to manipulate their singing and nearly any other sound.

For instance, Imogen Heap has demonstrated on numerous occasions the ability to "capture" a note between her thumb and forefinger. She can then stop singing but still have the note continue as long as she holds her fingers together. What's more, she can then raise or lower the pitch of the note by physically raising or lowering her hand. Moving her arm downward slows the speed of recorded sounds, opening her hand may release the note, and moving her arms in yet another way may change the type of effect currently used. These gestures are just a small example of what is possible, though. The sounds and effects produced by these gloves are completely programmable and only limited by the imagination of the user.

Even though these gloves are just making their public debut, performances by Imogen Heap demonstrate that new musical possibilities resulting from these gloves are quite impressive. Furthermore, this device has opened the door to even more innovative sound manipulation techniques. For example, Imogen Heap has combined these gloves with a Microsoft Kinect device, which detects where she is on stage and how her entire body is positioned at any given moment. Moving her hands one way while standing on one particular area of the stage produces one sound, while performing the exact same motion on another part of the stage yields a completely different sound. This extra level of interaction allows Imogen Heap, or potentially any musician, to change the sound of their music by physically moving

119

around on stage. This creates an enormous amount of potential to produce an interesting show that not only entertains the ears, but the eyes as well.

The revolutionary nature of these gloves may not be entirely apparent without seeing and hearing them for yourself. Therefore, as usual, I recommend a quick search online for a demonstration[64]. After which, I trust you will have a much better understanding of this musical innovation's capabilities. Furthermore, by considering how this type of technology may eventually be used in other areas of the music industry, we can start to derive a clearer picture of where musical creation, production, and performance in general may be headed.

For instance, it's not difficult to imagine these gloves used in combination with other instruments. A drummer could use hand gestures to change the sound of his drum kit in the middle of song. A guitarist could "pull" notes from his guitar that would continue to be played in

---

[64] Imogen Heap presented at Wired 2012 and provided a live demonstration of these gloves which has since been put online:

[64]Cornish, David. "Watch Imogen Heap's full Wired 2012 glove demo and performance (Wired UK)." *Wired.co.uk – Future Science, Culture & Technology News and Reviews (Wired UK)*. N.p., n.d. Web. 2 Sept. 2013. <http://www.wired.co.uk/news/archive/2012-10/29/imogen-heap>.

the background while he soloed over top of them. A group of musicians could capture and manipulate the sounds coming out of each other's instruments, which would transcend the level of collaboration that musicians have ever been able to achieve during a live performance.

As an example, the leader singer of a rock band could capture a note from the lead guitarist and hold it up to his mouth. At this point, when he continued to sing, the notes created by his voice would be transitioned to the higher octave of the note he "stole" from the guitarist. But this is just the tip of the iceberg.

A truly skilled musician may even be able to program a range of notes into the gloves and then play an invisible instrument on stage.[65] While this particular use of the gloves may be considered more of a novelty than truly groundbreaking, the thought of a musician

---

[65] The Awesome Foundation is already working on invisible instruments. This work isn't based on the "musical gloves," but is nonetheless similarly expanding how artists can create and perform their music. See the footnote below for more!

[65]"Invisible Instruments Rok! | The Awesome Foundation." *The Awesome Foundation*. N.p., n.d. Web. 2 Sept. 2013.
<http://blog.awesomefoundation.org/2010/11/09/invisible-instruments-rok/>.

creating music on an invisible piano, violin, or drum set does seem rather intriguing. What's more, all of these potential innovations are only scratching the surface of what may ultimately be possible in this area. Using gestures and digital music equipment to interact with sounds opens up more possibilities than could ever be written down. A small sample of these possibilities has been provided here, and it's my hope that interested readers will continue exploring this area further. Who knows, it may even be a reader of this book who popularizes some of these ideas in mainstream music someday.

The opportunities presented by the so-called "musical gloves" are but one example of how music creation, and music in general, may evolve. Whether or not this particular advancement will ever be fully realized is an uncertainty at this point. Nevertheless, the mere attempt to develop such a device, along with other innovations described in this section, illuminate the path that technological advancement is paving for the next generation of musical innovation. Creating and combining new sounds in novel ways through the use of technology is arguably the area of musical production that is most ripe for growth. As such, I would fully expect more devices, artists, and sub-genres to emerge dedicated to this budding field.

The music industry as a whole is in the midst of some rather profound changes. Distribution was turned on its head a decade ago with the introduction of file sharing. This transformation carries onward as music streaming and on-demand services gain popularity. The CD, like the cassette tape and 8-track before it, is losing popularity and beginning to show its age. Unlike its predecessors, however, there are no obvious replacements for the CD as the primary physical storage medium for audio. Unless a new medium emerges unexpectedly in the near future, this could mean we're witnessing the end of standardized physical music storage and the beginning of a purely digital music era.

Music creation and production are both reaching new heights thanks to the near-constant innovation occurring in computers, electronics, and digital audio equipment technologies. Furthermore, there are no signs of stagnation occurring in this area in the near future. In fact, due to some exciting developments, such as the digital guitar and the musical gloves, it wouldn't be

123

surprising to see some of the most technologically innovative music ever created released in the next few years.

The entire music industry is being renewed, reshaped, and reborn before our very eyes -- not to mention our ears. It's not been an easy transition, but if the innovations that we're seeing today are a sign of what's to come, then the industry's brightest days may still lie ahead. As a technology enthusiast and a huge music fan, this is one area I'm particularly excited about and will be monitoring closely.

Discussion Questions:

1) Do you agree with the author's assertion that file sharing has had the largest impact on the state of the music industry over the past decade? Why or why not?

2) Compare and contrast traditional (terrestrial) and Internet-based radio. Which of these delivery methods do you prefer and why?

3) Will Internet-based radio ever fully replace its conventional counterpart? Why or why not? Explain.

4) Will there be a physical medium for music storage that replaces the CD? If so, describe its features and abilities. If not, how do you envision music will be enjoyed in the car, while working out, and while on the go?

5) Are record labels becoming less relevant to the success of an artist? In the future, will record labels continue to be the primary way that artists achieve success? Explain.

6) Will we ever see a day where record labels disappear from the music industry entirely? Why or why not?

7) Do you agree with the idea that electronic music makes use of new technologies more frequently than other genres? Explain.

8) Do you think we will see more innovation in the genre of electronic music than in other genres in the next decade? Why or why not?

9) What future do you envision for devices like digital music gloves? Will this type of "instrument" catch on and inspire other novel digital-based instruments or is this simply a fad? Why do you think this is the case?

Chapter 6: Video Games

From Fringe Hobby to Big Business

It's doubtful that many people realized the significance of the Magnavox Odyssey game console when it was released to the public in 1972.[66] At that time, society was still just getting used to the newly-popularized color television. The idea of interacting with virtual objects on a television set was both largely unheard of and difficult to grasp for many. These factors likely helped shape the fate of the hardware, which wasn't a runaway success with the public. Nevertheless, this device paved the way for an entirely new industry. An industry that wouldn't only shape the entertainment of future generations, but launch a sizable sub-culture, a growing field of career opportunities, and an ever-increasing list of exciting experiences that are *only* made possible by video games.

Today, the video game industry is estimated to be worth around $80 billion, a value that continues to climb each year.[67] In recent years, a large part of that growth is

[66] "Ralph Baer: Recovering the History of the Video Game." *Smithsonian*. N.p., n.d. Web. 7 Sept. 2013.
<invention.smithsonian.org/resources/online_articles_detail.aspx?id=531>.

[67] Video games: Battle of the boxes | The Economist." *The Economist - World*

owed to a number of fairly radical developments currently taking place. Many of these developments stem from larger movements in the entertainment industry as a whole, such as the push toward "social" entertainment. However, there are a handful of interesting trends occurring today and on the horizon that are specific to gaming. These trends will be the primary focus of this chapter, but we'll begin with a brief look at some of the largest and most profound developments currently impacting the video game industry.

Social Gaming

One of the largest overall trends in the entertainment industry right now is the use of a "social" component. This idea was discussed in some depth earlier in the sections dealing with movies and television, but it applies equally to the topic of gaming. In fact, since gaming is an interactive experience by definition, the idea

---

News, Politics, Economics, Business & Finance. N.p., n.d. Web. 7 Sept. 2013. <http://www.economist.com/news/business/21578427-microsofts-newest-games-console-has-entire-living-room-its-sights-battle-boxes>.

[67] "Video Game Industry Set for Growth -- Global Game Market Forecasted to Grow to $82 Billion Five Star Equities Provides Stock Research on Sony and Majesco Entertainment" Yahoo! Finance - Business Finance, Stock Market, Quotes, News. N.p., n.d. Web. 7 Sept. 2013. <http://finance.yahoo.com/news/video-game-industry-set-growth-122000072.html>.

of creating a social experience by including multiple participants is much more intuitive in this space.

It should be noted that even though games are becoming social in increasingly novel ways due to the rise of the Internet and mobile devices, the idea of "social gaming" isn't anything groundbreaking. In fact, today's "social gaming" typically looks a lot like what's been called "multiplayer gaming" in the past. That is, the primary way to make a game "social" is to allow multiple participants to play together. In the past, meeting with friends in one room and plugging in multiple controllers to a console accomplished this. Of course this still happens, but today, the vast majority of multiplayer (or "social") gaming takes place via the Internet.

Online gaming typically involves avatars that interact in some way, as well as a communication mechanism whereby players chat via text or voice. Social interaction is usually encouraged or even required throughout said games. Of course, there are many variations of this and different twists on these ideas that allow for novel types of gameplay. Throughout this section, we'll discuss many of these social gaming developments as they relate to other exciting innovations in the field.

Gaming in 3D

Another major entertainment industry-wide trend that's impacting gaming is the use of stereoscopic 3D. This technology allows viewers to perceive the image as if it were "jumping out" of the screen.[68] As described previously, 3D films are making waves in the motion picture industry, while struggling quite considerably to catch on in other areas: television and video games.

To be sure, the debate still rages between gamers and others involved with the industry over whether or not the use of 3D in video games has been a success. Some gamers will undoubtedly argue that 3D technology achieves realism that just isn't possible otherwise. Meanwhile, critics of the technology counter that the extra equipment, costs, and discomfort associated with 3D games outweigh the benefits.

Although both sides of this debate have merit, few would disagree with the assertion that the use of 3D in video games has not swept the industry by storm. It has undoubtedly been a noticeable trend in recent years,

---

[68] If you need an example, look no further than the Nintendo 3DS.

but the use of 3D in gaming is still far from being standard practice. Furthermore, whether or not this trend will have any major impact on the future of gaming is, at this point, still up for debate.

Nonetheless, 3D games will likely continue to exist as a niche market, since sales demonstrate that there is at least *some* demand for these games. It's also possible that we'll see the popularity of 3D games rise as other futuristic gaming technologies are introduced.

For example, one idea just on the horizon is the virtual reality headset. This type of device essentially requires compatible 3D games in order to deliver the proper experience. If this type of gadget were to catch on with the public, it's foreseeable that the popularity of 3D games would increase, as well. Of course, this is only one example. Nevertheless, it does illustrate that the success or failure of 3D video games may ultimately depend on the public reception of other emerging innovations in the field. We'll explore the potential of such future technologies in much greater depth a bit later in this section.

Second Screen Technologies in Gaming

Yet another industry-wide trend that's affecting the world of video games is the "second screen" or "companion device." As discussed earlier, the use of a mobile device to enhance the movie or television experiences has been increasing in popularity. It should not come as much of a surprise that this technology is also being tried in the realm of video games.

Interestingly, the use of a second screen to improve the video gaming experience actually predates the use of the technology in other entertainment fields.

The technique was first attempted, to a certain degree, back in 1998, with the release of the Sega Dreamcast console. More specifically, the Dreamcast came equipped with a removable peripheral device known as the "Visual Memory Unit" (VMU). This small device could be inserted in controllers and act as a standard memory unit *or* it could be used as a second method of interacting with the game. The device was rather small in size, but it did have five buttons and a screen. These features made it capable of not only receiving input from the user, but displaying information as well. While gaming on the console, the user could glance down at the device and receive extra information about what was occurring in the game. In other words, it truly was a functional, albeit primitive "second screen" device.

Today's second screen technology has improved on this early forerunner quite considerably. For starters, the size of the typical second screen has increased from the Dreamcast VMU's tiny dimensions (1.46 in x 1.02 in) to at least three times this size, but usually even larger.[69] Most commonly, the second screen experience tends to be driven by an app running on a tablet or other mobile device, so the dimensions will vary based on the hardware of choice.

It's no surprise that the quality of mobile devices have been steadily improving, which has allowed the potential quality of the second screen experiences possible to improve as well. Today, many mobile devices sport top-notch processors, high-quality displays, and all-around stunning graphics capabilities. All of this results in the potential for a second screen experience that's a far

---

[69] "Visual Memory Unit." *Sega Retro*. N.p., n.d. Web. 14 Sept. 2013. <segaretro.org/Visual_Memory_Unit>.

cry from the black and white VMU's available a decade ago.

As far as usability goes, the use of a second screen during gaming is not all that different from how the technology is used in other forms of entertainment.

Second screen gaming experiences provide additional information to gamers that can be viewed at a glance, without interrupting what's displayed on the main screen. Typically, this includes world maps, inventory lists, character statistics, or any other information that the gamer might need at a moment's notice. However, the second screen can go far beyond these relatively minor enhancements to provide alternative views of on-screen action, live video chatting capabilities, and other more substantial supplementary content.

For example, a list of nearby characters and items could be populated as you wander through a virtual world. Or, in-game action from the enemy's point of view could be displayed on the second screen during intense moments of battle. Taking this idea a bit further, entire games could be built around the technology by displaying one virtual environment on one screen, while simultaneously showing a different one on the other.

As an illustration of this idea, consider a game dealing with time travel that had second screen capabilities. Gameplay could show the modern-day world on the main screen, while showing the exact location as it appeared in the past on the second screen. As the character walks around, the view of both time periods would change. Time travelling from one period to the other would cause displays to switch, resulting in the modern day world to appear on the second screen, with the historical world now on the main screen.

Of course, this is but one idea. This concept would work equally well with a large number of other ideas including games built around ghosts, creatures, or any other item that could only be seen on the second screen.

At the end of the day, since technologies put into tablets and other mobile devices have improved so greatly, the use of the second screen to enhance gaming is only limited by the imaginations of game developers and the demand for such experiences by gamers. Interestingly enough, the reception of such technologies hasn't been overwhelmingly positive.

At present, there's an ongoing debate occurring

within the gaming community over whether or not the use of second screen tech is the "future" of gaming or just a passing trend. In late 2012, Nintendo made a rather large gamble that gamers would be in favor of a console centered on the second screen, with the release of the Wii U. This console featured a touchscreen embedded in its primary game controller. Over a year later, it's pretty apparent that the console is struggling to catch. Sales figures aren't impressive and a fairly large segment of gamers don't seem interested in the device.[70]

To be fair, there are many possible reasons why the Wii U has been received poorly. The "second screen" feature *may* have contributed to this poor reception, but it's also possible that sales would have been even worse if the controller didn't have a built-in screen. Nevertheless, since the Wii U features the second screen technology so prominently, the fate of the feature seems to be tied to that of the console in the eyes of many

---

[70] According to Nintendo's own data, the Wii U has sold approximately 3.61 million units as of June 2013:

[70]"Consolidated Sales Transition by Region." *Nintendo Co., Ltd.*. N.p., n.d. Web. 14 Sept. 2013. <www.nintendo.co.jp/ir/library/historical_data/pdf/consolidated_sales_e13 06.pdf>.

gamers. Poor sales signal to many that second screen technology has failed to take hold in the gaming community. However, in reality, the role of the second screen in the world of gaming goes far beyond the Wii U.

The history of gaming is considerably more intertwined with the concept of the second screen than perhaps most realize. Ever since the VMU was released by Sega for their ill-fated Dreamcast console, numerous other second screen devices have cropped up.

The original Sony PlayStation allowed gamers to connect a device known as the "PocketStation" for use as a second screen back in 1999. Three years later, in 2002, the Nintendo GameCube console offered gamers the ability to use a Game Boy Advance as a controller, effectively turning the portable device into a second screen for the GameCube.[71]

In 2006, the original Nintendo Wii was released with support for connectivity to Nintendo's handheld console of that time period, the Nintendo DS. Also in

---

[71] "Miyamoto Wants To Return To Unrealised GameCube-GBA Link Ideas With Wii U - Wii U News @ Nintendo Life." *Wii U, 3DS & eShop - News, Reviews & Forum - Nintendo Life*. N.p., n.d. Web. 15 Sept. 2013. <http://www.nintendolife.com/news/2012/12/miyamoto_wants_to_return_to_unrealised_gamecube_gba_link_ideas_with_wii_u>.

2006, the initial launch of the Sony PlayStation 3 brought capabilities for gamers to connect their PlayStation Portable (PSP) devices via Sony's Remote Play technology. The other consoles in this generation also supported connectivity to external devices in some capacity.

As mentioned, Nintendo's Wii U relies heavily on the feature. Sony adapted their Remote Play technology to allow for the connectivity of their next-generation products: the PlayStation 4 and PS Vita. In addition, Sony also introduced the ability for customers to use their iOS and Android devices to retrieve data related to their PlayStation Network accounts and activity. Not to be outdone, Microsoft announced Xbox SmartGlass in 2012, which allows iOS, Android, and Windows devices to communicate with Microsoft's Xbox 360 and Xbox One consoles directly, opening the door for true second screen gameplay experiences on both consoles.

Just about every major console released in the past decade has offered gamers the chance to use a second screen, in one form or another. The Wii U may have been the first console to feature this technology as its centerpiece, but the tech is already quite ingrained in the gaming community.

With this in mind, regardless of how the Wii U fares, second screen technology will likely continue proliferating. There certainly are many gamers that have no interest in games that utilize a second screen, but there certainly are others who believe the technology offers innovative, unique, and exciting experiences that just can't be obtained via traditional gameplay. It's possible that second-screen gaming will never overtake traditional single-screen gaming as the play method of choice, but it seems likely that second screen technology will always be around in some form within the realm of gaming.

The Great Transition to Digital

A larger trend that's impacting the video game industry is the transition from physical to digital storage. For the same reasons that are causing this trend in the music industry, there's been a noticeable shift away from physical media in recent years. Instead, video games are increasingly being offered as digital downloads, which are then saved directly to consoles and PCs. This essentially eliminates the need for discs or cartridges.

Companies behind the production and distribution of video games are pushing for this transition primarily to save on packaging and distribution costs, but there are also some advantages from the consumer's point of view. While the PC gaming community as a whole has largely supported the shift towards digitally-distributed games, the console-based gaming community has remained quite skeptical. Nevertheless, the steady influx of online game distribution services like Steam, Direct2Drive, and OnLive point to a future ruled by digital. When factoring in that a large portion of gaming now occurs on mobile devices, which retrieve their content digitally from mobile app stores, it becomes increasingly likely that the days of video games sold on physical discs are numbered.

For readers who aren't involved in the gaming community, it's important to note how much controversy surrounds the push for digital game distribution. One of the primary concerns from gamers is that purely digital games raise a number of questions related to secondhand sales.

For instance, gamers have raised questions about whether they'll be able to sell a game after they are finished with it, who they will they be able to sell it to, and if the value of used games remain the same as their physical counterparts today. These questions are only a small portion of the concerns being expressed by gamers about the coming shift to digital games, but the overriding message is clear: this transition is not as straightforward as it initially appears, and game companies will need to work very closely with the gaming community to ensure the benefits of digital gaming are felt by everyone.

Gamification's Impact on Gaming

A final trend that's been impacting the entire entertainment industry to a certain degree over the past few years is a topic that was already discussed quite a bit: gamification.

As described earlier, gamification is a term that refers to the process of adding elements of gameplay to a non-gaming experience. For example, the Nissan Leaf automobile encourages drivers to save energy by calculating the fuel efficiency of their driving and reporting this number to drivers using little pine tree icons.[72] Drivers can then post this information online to compare how their tree count ranks among other drivers, how their country's tree count stacks up against other countries, and add to the total number of trees saved by drivers of the Nissan Leaf worldwide.[73] This feature has created a competition for drivers to get the highest number of trees possible, but in doing so, they are also reducing carbon dioxide emissions. This particular example would be considered a form of environmental gamification, but the methodology is applicable to just about anything. Tallying up daily physical activity, doing chores, or even budgeting money are all potential candidates for gamification.

As the trend of continues to spread, its impact on the video game industry may start to become more noticeable. That is, there is currently a pretty large divide between the quality of gamification "games" and mainstream video games, with video games coming out

---

[72] "Gamification of Environment | Gamification.org." *Gamification Wiki*. N.p., n.d. Web. 22 Sept. 2013.
<http://gamification.org/wiki/Gamification_of_Environment>.

[73] "You+Nissan ." *Nissan cars, vans, fleet and services*. N.p., n.d. Web. 22 Sept. 2013.
<http://www.nissan.co.uk/GB/en/YouPlus/welcome_pack_leaf/eco_virtual_trees.html>.

on top in nearly every way. However, in recent years there have been a number of fitness games released for the major consoles, which have started to close this gap. The ultimate goal of these games is to assist the player in achieving their fitness goals, but the entertainment value is quite apparent as well. This type of game points to a future where high-quality video games assist people with real-world tasks and goals. As more areas of society embrace this trend, it's likely that we'll see other video games created in a similar vein, but with different goals.

The Future of Gaming Consoles

It's no secret that the role of gaming consoles is changing. While at one time consoles were purchased almost exclusively for their game-playing abilities, today's consoles expand far beyond the traditional offerings to provide a more complete entertainment experience. This has resulted in a shift away from the modern gaming console being a sole-purpose device entirely focused on playing video games, to a multi-purpose streaming video receiver, media player, web browser, *and* a gaming machine.

This strategy has been a bit more of a gamble

than some readers may realize, as it's risked alienating some gamers who may believe that a console should only focus on gaming. Nevertheless, it appears that this strategy has allowed consoles to not only remain relevant in some fairly turbulent times for the industry, but also continue as a successful and highly profitable entertainment option, by nearly any measurement.

This success in extending the capabilities of consoles to other areas of entertainment has motivated the designers of next-gen consoles to continue offering these features and even expand into new territory.

As an example, all of the major latest-generation consoles are continuing to offer the ability to connect to Netflix, Hulu, and other Internet-based streaming services. Meanwhile, Microsoft's Xbox One console is, for the first time, offering gamers the ability to connect a cable box to the device and use the console as a front-end for their cable or satellite subscription.[74] This may seem like a fairly insignificant development, but each step in this direction is another step towards a takeover by gaming consoles as the go-to centerpiece of the home

---

[74] "Get the facts - Xbox.com." *Xbox - Xbox.com*. N.p., n.d. Web. 23 Sept. 2013. <http://www.xbox.com/en-US/xbox-one/get-the-facts#6-5>.

entertainment system.

The idea of having one device that can "do it all" does have a certain amount of appeal. Changing television inputs, keeping track of multiple remote controls, managing numerous power adapters, wrangling unruly wires, and finding storage space for a half-dozen devices are just a few of the headaches that come with using more than one device. Furthermore, it makes sense that consoles are branching out toward other entertainment areas, rather than the reverse, since consoles already -- by necessity -- have some pretty impressive technical specs. That is, it would be extraordinarily difficult for cable companies to suddenly start offering video games for their cable receivers, as they often don't have the necessary hardware. However, this doesn't mean that major game consoles are going uncontested in the realm of video gaming on the family television. In fact, recently, the gaming console has seen some new and very interesting challengers in this space.

One notable newcomer to the world of gaming consoles appeared on the scene in the summer of 2009:

the cloud-based OnLive gaming console.[75] Released with a great deal of media buzz and excitement, the modestly-priced console promised gamers the ability to always be able to play the latest and greatest games without having to upgrade the hardware. This was made possible by performing all of the necessary computations on industrial grade servers in the cloud, while the user simply receives the resulting video feed on their television via the console back at home.[76] The idea was that the servers could be upgraded to support the latest games, while the customer continued using the same OnLive console to connect to the servers. In this way, the user could theoretically use the same OnLive console for many years, while also being able to play the newest games.

At the time, this was somewhat of a revolutionary concept that seemed to be destined for success. However, the service never quite lived up to

---

[75] "Interview: OnLive CEO Steve Perlman gives us his post-launch perspective | Crave - CNET." *Technology News - CNET News*. N.p., n.d. Web. 23 Sept. 2013. <http://news.cnet.com/8301-17938_105-20010687-1.html>.

[76] Grant, Christopher. "GDC09 interview: OnLive founder Steve Perlman wants you to be skeptical | Joystiq." *Joystiq*. N.p., n.d. Web. 23 Sept. 2013. <http://www.joystiq.com/2009/04/01/gdc09-interview-onlive-founder-steve-perlman-wants-you-to-be-sk/>.

expectations. Amongst other issues, gameplay was entirely reliant upon the cloud, which meant that any hiccups in network connectivity resulted in a significant hit on performance and usability. The console and accompanying cloud service are still around, but the console has been in the news more often lately for turmoil occurring within the company than for the console or services it provides.

In August of 2012 the company laid off all of its employees and announced the CEO, Steve Perlman, was stepping down.[77] There's always a chance that OnLive could make a comeback and rise up as a competitor to the mainstream consoles, but at this point, all signs point toward OnLive being little more than a great idea that just didn't work out as planned.

Another competitor to top-tier gaming devices that garnered quite a bit of attention in the summer of 2012 was the OUYA console. This tiny device, dubbed a "micro-console" by some, was created by a small startup

---

[77] "OnLive's Steve Perlman bids farewell, says other projects will 'blow your mind'."*VentureBeat | Tech. People. Money.*. N.p., n.d. Web. 23 Sept. 2013. <http://venturebeat.com/2012/08/28/onlives-steve-perlman-says-farewell-says-other-projects-will-blow-your-mind/>.

and crowd-funded through a Kickstarter campaign.[78] OUYA's most innovative aspect was its hardware, which was designed to be modifiable and as open source as possible. OUYA runs an Android-based operating system, which meant it would already have a fairly sizable catalogue of Android games available by the time it launched. In addition, all consoles contained the necessary components to allow for game development. This meant that anyone could develop games for the console without paying additional licensing or development fees. Unsurprisingly, the OUYA broke records for the most money ever raised via Kickstarter in a 24-hour period, and made headlines across the world for its innovative and potentially disruptive nature.

In the summer of 2013, OUYA was released to the general public. Unfortunately, the implementation of the console could not quite live up to the vision, as the device was met with largely negative reviews.[79] Nevertheless,

---

[78]Kickstarter is a web-based service that helps people gain funding for their projects. Please see the website for more: http://www.kickstarter.com/hello

[78]"OUYA video game console becomes the latest crowdfunding success." *TECHi*. N.p., n.d. Web. 23 Sept. 2013. <www.techi.com/2013/06/ouya-video-game-console-becomes-the-latest-crowdfunding-success/>.

[79] "Ouya Review - IGN." *Video Games, Wikis, Cheats, Walkthroughs, Reviews,*

even if OUYA doesn't ultimately turn out to be the disruptive innovation that many thought it would, the device and its immensely popular crowd funding campaign have shined a light on what gamers want to see in future consoles.

The OUYA may have failed by some accounts, but the successful fundraising has nonetheless proven that the console's features are both popular and potentially profitable. With this in mind, it seems highly likely that these ideas will reappear, either in the form of another startup or perhaps as a new direction for one of the industry's major players. Regardless of how these ideas are ultimately implemented or who puts them into action, it'll be interesting to see these concepts fully realized.

In short, the OUYA and OnLive gaming systems were both highly anticipated devices that, by many

News & Videos - IGN. N.p., n.d. Web. 29 Sept. 2013.
<http://www.ign.com/articles/2013/07/26/ouya-review>.

[79]"5 Ways the Ouya Game Console Failed - Ouya's woes." *Tom's Guide: Your High-Tech Source of Information*. N.p., n.d. Web. 29 Sept. 2013.
<http://www.tomsguide.com/us/ways-ouya-console-failed,review-1881.html>.

[79]"OUYA Gaming Console Review." *LAPTOP Magazine - Product reviews, tech news, buying guides, and more*. N.p., n.d. Web. 29 Sept. 2013.
<http://www.laptopmag.com/reviews/consoles/ouya-gaming-console.aspx>.

accounts, failed to live up to expectations. Each promised some fairly innovative features, many of which had not yet been seen in the industry until the launch of these consoles. Nevertheless, the challenges facing these newcomers proved to be far too daunting. To anyone familiar with the industry, the difficulties faced by these up-and-comers didn't come as a surprise since the competition in this space are console titans such as Microsoft, Sony, and Nintendo -- each of which has spent years building the enormous infrastructure and customer base necessary to be successful. At the end of the day, enticing potential customers to take a chance on a brand new gaming device when there are already so many well-received and far more capable alternatives is quite simply an exceedingly difficult task.

While the console market seems to be somewhat resistant to newcomers, innovation, and change, the much younger and more nimble *mobile* gaming space has been thriving on these qualities in recent years.

The Role of Mobile in Gaming

Another area of growth that almost no one will be surprised to see more of is mobile gaming. The term "mobile gaming" has been around at least as long as the Game Boy, but it's taken on an entirely new meaning in recent years as smartphones, tablets, and mobile apps have risen dramatically in popularity. The number of games played on these devices has skyrocketed, and by some measurements, this type of gaming is outpacing the growth of traditional consoles.[80]

With that in mind, the role of mobile devices can't be ignored when determining where the gaming industry may be headed. It's a near certainty that mobile gaming will have an impact on the success of future. What this impact will be and how exactly it'll affect the future of gaming, however, is still very much up for debate.

In fact, there's a considerable amount of uncertainty surrounding the current state of mobile gaming and where it may be headed in the near term.

---

[80] "App Annie & IDC Portable Gaming Report Q2 2013: iOS & Google Play Game Revenue 4x Higher Than Gaming-Optimized Handhelds." *App Annie Blog | Apps. Stats. Insights.*. N.p., n.d. Web. 22 Sept. 2013. <http://blog.appannie.com/app-annie-idc-portable-gaming-report-2013-q2/>.

As discussed earlier, the mobile gaming industry has been around *at least* since the release of the original Nintendo Gameboy circa 1989. However, today's mobile gaming industry is no longer centered on Nintendo or any other sole-purpose device built exclusively for gaming. Rather, the majority of growth in the mobile gaming market today is found in the ever-growing smartphone and tablet arena.[81] Perhaps unsurprisingly, the recent explosive growth in this space has led to quite a few interesting and exciting innovations that may eventually unseat the traditional console as king of the gaming market.

The beginning of this growth can be traced to 2007, with the release of the original iPhone.

There were plenty of other smartphones on the market prior to the iPhone's release, but there's no doubt that Apple's handset catapulted this category of devices to a whole new level in terms of capability and features.[82] After App Store's unveiling in 2008, all of the

---

[81] "iOS & Google Play Game Revenue 4x Higher Than Gaming-Optimized Handhelds." *App Annie*. N.p., n.d. Web. 5 Oct. 2013. <inbound.appannie.com/app-annie-idc-gaming-report-q2-2013>.

[82] The exact definition of "smartphone" leads to some contention in this

pieces were in place for gaming on mobile devices to explode in popularity, which is exactly what's occurred.

There are quite a few reasons why gaming's caught on so quickly in the mobile space, but there are a handful of factors that played a vital role in that success.

First, mobile gaming is extraordinarily convenient. Since many people already carry around a smartphone nearly everywhere, it makes sense that they'll play a game while waiting in a lobby, on a bus, or anywhere else they may find themselves with some time to kill.

Second, the performance of smartphones has improved enough over the past few years to allow for stunning graphics and smooth gameplay experiences. Make no mistake, these are still mobile devices that can't quite compete with the latest gaming console or a high-end PC, but the graphics on mobile devices are nevertheless respectable for the space they're filling.

Third, games for mobile devices are well known for being cheap. In fact, many mobile games are available for download completely free of charge. Some of these

---

area. However, if we define a smartphone as a handset that allows users to do more than simply make calls and send text messages, then smartphones most certainly predate the iPhone.

games offer extra content that can be purchased, while others include advertisements to turn a profit. In fact, other games require users to purchase a "full version" to experience all they have to offer. Even still, these purchases are often far less expensive than the asking price of a typical console or PC game.

Finally, mobile devices offer innovations and features that just aren't possible on a traditional console.

For example, the touch screen is a staple in the modern smartphone design. As such, most games use the touchscreen as the primary form of interaction. *Angry Birds* and *Fruit Ninja* are just two of the touchscreen gaming kings to capitalize on such a design.

In *Angry Birds*, players fling fowl into structures occupied by their arch enemies: pigs. In *Fruit Ninja*, gamers are tasked with slicing and dicing fruit to earn points.

When considering the simplicity of these games, it may be tempting to wonder how they became bestsellers. However, upon playing these games or speaking with someone who has, it quickly becomes apparent that the primary draw is the addictive experience, driven by the interaction with the touchscreen device.[83]

Slicing through fruit via a joystick or controller likely wouldn't result in an equally-appealing experience since players would presumably have to guide a reticle or some other on-screen icon to cut the appropriate spot. This would completely alter the gameplay experience in a way that would likely be far less appealing to most users. *Angry Birds*, and countless other mobile games, suffer the same fate if the touchscreen interaction is replaced with a traditional joystick or controller. The touchscreen is one of the linchpins of the modern mobile gaming experience.

Social gaming is another area where mobile gaming is thriving. In fact, some of the most successful mobile games of all time have fallen under the umbrella of "social," including two of the best-selling mobile apps

[83] "Fruit Ninja is the best-selling Windows Phone 7 game | VG247." *VG247 | VG247.com*. N.p., n.d. Web. 6 Oct. 2013.
<http://www.vg247.com/2011/01/09/fruit-ninja-is-the-best-selling-windows-phone-7-game/>.

[83] "Angry Birds is the App Store's best-selling game ever - VideoGamer.com."*PC and Console*

[83] *Game News, Videos, Reviews - VideoGamer.com*. N.p., n.d. Web. 6 Oct. 2013.
<http://www.videogamer.com/news/angry_birds_is_the_app_stores_best-selling_game_ever.html>.

to date: *Words With Friends* and *Draw Something*.[84] Of course, console or desktop-based games are just as capable of succeeding in the social arena, and there are indeed many of such success stories. However, social games designed for mobile devices have been able to leverage the fact that people tend to have their phones with them throughout most of the day. This attribute of modern society has allowed mobile developers to create interesting gameplay paradigms.

For example, one popular gameplay experience allows folks to play one or more ongoing games spanning an entire day, multiple days, or even a few weeks at a time, playing at their leisure. Both *Words With Friends* and *Draw Something* used this format to great success. It should be noted that the plausibility of this type of gameplay succeeding on a traditional console is somewhat doubtful since such hardware is simply not accessible throughout the day in the same way as a mobile device. This seamless integration of social and mobile is yet another area of the gaming market where

---

[84] "The Best-Selling iPhone Apps - Business Insider." *Business Insider*. N.p., n.d. Web. 6 Oct. 2013. <http://www.businessinsider.com/the-best-selling-iphone-apps-2013-5?op=1>.

mobile endeavors have some clear advantages over consoles.

Another area where mobile gaming has a leg up on traditional consoles is in the space of augmented reality (AR).

Augmented reality is the concept of layering digitally created objects onto one's surroundings to create an environment partly in the physical and virtual worlds.[85] This technology will be discussed in more depth later, but as an introduction, consider the following example.

A game based on AR tech might require a player to physically travel to a location in the real world, and upon arrival, the player would look around at the location using their mobile device's camera, which would superimpose digital in-game objects onto the device's screen. The player could then interact with these objects through their device. A mundane street sign could reveal important in-game clues when looked at through a smartphone. Perhaps other players would be recognized by the game and be replaced with an avatar when viewed

---

[85] Chapter four's section on the future of health-related technologies contains a discussion of augmented reality.

on screen. Of course, these are just a few of the nearly infinite possibilities created by this type of technology.

Considering that the core attribute of augmented reality gaming is viewing the physical world through the eye of a camera, it's easy to see why this type of gaming would be at a disadvantage on a traditional console, which is stationary by nature. At present, this lack of capability regarding augmented reality gaming isn't affecting the console market, as this type of gameplay is still in its infancy. However, as the technology improves and games come closer to reaching their potential, augmented reality gaming may become yet another innovation that helps mobile devices overtake consoles as the primary gaming platform.

Even at present, with augmented reality games perhaps still years away from becoming a mainstay, the growth of mobile gaming has been highly impressive. Considering the platform was nearly non-existent just a few years ago, some readers may be surprised to learn that mobile gaming apps earned roughly $8 billion in 2012, with that number expected to climb.[86] Meanwhile,

---

[86] White, Martha C.. "Game Over? Why Video-Game-Console Sales Are

console sales declined 27 percent in 2012 compared to 2011.[87]

Furthermore, a recent poll of game developers by *The Game Developers Conference* revealed that a whopping 58 percent of game developers plan on developing their next game for a mobile platform.[88] With these stats in mind, it's becoming apparent that if consoles hope to compete with mobile devices, they must innovate and offer unparalleled gaming experiences that can't be matched.

Gesture-based Gaming

A popular innovation that has appeared in one form or another on all three of the major consoles in recent years is gesture-based interaction. Although each console manufacturer implemented this feature in their

Plummeting | TIME.com." *Business & Money | The latest news and commentary on the economy, the markets, and business | TIME.com*. N.p., n.d. Web. 7 Oct. 2013. <http://business.time.com/2013/02/11/game-over-why-video-game-console-sales-are-plummeting/>.

[87] "NPD report finds Xbox 360 'dominated' 2012 console sales, 890,000 Wii Us sold in the US so far." *Engadget*. N.p., n.d. Web. 7 Oct. 2013. <http://www.engadget.com/2013/01/10/npd-report-finds-xbox-360-dominated-2012-console-sales-890-00/>.

[88] "Game Consoles Are Already Dead — And Developers Know It." *Web Apps, Web Technology Trends, Social Networking and Social Media – ReadWrite*. N.p., n.d. Web. 7 Oct. 2013. <http://readwrite.com/2013/03/04/game-consoles-already-dead-developers-know-it#awesm=~ojypSWt9u3euoe>.

own way, offering a slightly different experience, the core principles are roughly the same. Essentially, players interact with the device using some sort of gesture or hand movement. Depending on the hardware, users either need a camera fixed on them, a motion-sensing controller in hand, or both.

Most readers will likely be familiar with this type of gameplay, as a motion-sensing controller is the primary form of interaction for the Wii console, which has been around since 2006.[89] However, the technology behind this type of interactive gameplay hasn't remained stagnant since that time. In fact, the release of the Kinect device for the Xbox 360 revolutionized this segment in 2010 by allowing a player to interact with a game without any controller whatsoever. What is required, though, is that players remain in a relatively fixed location so the console's camera can detect their movements. Hence, this type of interaction is not as practical for mobile games, which are typically played while on the go.

However, it certainly would not be impossible to

---

[89] "Corporate Management Policy Briefing : Q&A." *Nintendo Co., Ltd.*. N.p., n.d. Web. 10 Oct. 2013.
<http://www.nintendo.co.jp/kessan/060607qa_e/index.html>.

port this type of technology to a mobile device. After all, these gadgets are getting more powerful and most already have cameras. There would, however, be a few significant roadblocks to using this type of interactive gameplay on mobile hardware.

First, mobile games are designed to be played anywhere, while waiting in the doctor's office, sitting on a bus during the morning commute, at a restaurant, etc. Performing odd-looking and perhaps somewhat embarrassing gestures in these types of public settings is something that would likely face some difficulty catching on.

Second, mobile devices are typically held at arms-length, in one hand. Since this distance is only far enough away for the camera to capture the upper torso and the user only has one free hand, this doesn't leave much of the user's body left to perform the interaction, which creates some clear limitations surrounding gameplay possibilities.

It should be noted that none of this is meant to suggest gesture-based gaming will not be successful on mobile platforms. On the contrary, since gesture-based gaming has already found success on consoles, the next logical step is for this type of gaming-mechanic to spread to mobile. Nevertheless, there are still some technical and societal challenges that must be overcome before this next step can be taken. In the meantime, gamers seeking gesture-based gameplay experiences will be much more likely to turn to a console. This near-exclusivity is exactly the type of feature that console makers need to be seeking out and embracing if they hope to continue competing with the cheaper and more convenient alternatives presented by mobile hardware.

IllumiRoom

Another promising feature that may fall under the umbrella of ideas that can help keep consoles relevant is an innovation first revealed by Microsoft in January of 2013 at the annual Consumer Electronics Show (CES).[90] This technology, dubbed "IllumiRoom" is another one of those inventions that simply must be seen to be truly appreciated. As such, readers are strongly encouraged to take a moment to watch a video or two showcasing the

---

[90] "Microsoft IllumiRoom concept demoed at CES... Xbox 720 anybody? (video) - Pocket-lint." *Pocket-lint - Gadget Reviews, Product News, Electronic Gadgets*. N.p., n.d. Web. 12 Oct. 2013. <http://www.pocket-lint.com/news/118863-microsoft-illumiroom-concept-demo-video-xbox-720>.

concept.

For readers who can't access videos at the moment, the idea behind IllumiRoom is to use a projector in concert with a television to create the illusion that gameplay elements are jumping out of the TV and into the living room. The television remains the primary focal point, but the area around the television is now used to display in-game elements in the player's peripheral vision. The effect that this setup produces is certainly unique, and its appealing nature is readily apparent after watching just a few short demonstrations. In addition, the fact that this fairly complex setup requires a television, a projector, and the user to remain in one fixed location while playing essentially guarantees that this type of technology will not be ported to a mobile device anytime soon. This is an extraordinarily promising innovation that may help consoles compete with mobile gadgets in the coming years.[91]

---

[91] Or at least in a future generation Microsoft Xbox, since the IllumiRoom is a Microsoft technology!

Virtual Reality Headsets

Yet another technology just on the horizon that would be difficult to use while on the go is a device that gamers have been wishing for since the dawn of the gaming: the virtual reality headset.

Virtual Reality (VR) is the concept of using technology to trick a user's senses into believing that they're physically located in another place, whether real or fictional. The primary sensory alteration associated with VR is vision, but the addition of audio to the experience is fairly common as well.

Various VR headsets have emerged in the past few decades only to disappear shortly after their introduction to the public. For the most part, this failure of VR headsets to catch on seems to be technical in nature, as opposed to the devices being rejected by the gaming community. That is, in past attempts, technology had not quite reached a point where a convincing experience could be created. However, recent events have demonstrated that not only is the demand for this type of device just as high as it has always been, but the technology (and its associated costs) may finally be ready to make a lasting impact on the gaming industry. The majority of these events revolve around a device that had a highly successful Kickstarter campaign in 2012: the Oculus Rift, which raised an impressive $2,437,429 from over 9,500 supporters.[92]

The reason why the Oculus Rift has been greeted with so much enthusiasm by the gaming community is the same as why this device offers so much promise for the future: it actually manages to deliver on the promise of a unique and engaging virtual reality experience. Whereas most other VR headsets have suffered from slow response times, headache-inducing graphical displays, or counter-intuitive controls, the Rift manages to capitalize on recent technological improvements in the mobile realm to provide a VR experience that's truly exceptional.

At present, the user experience for the Oculus Rift consists of strapping on a headset, which has two adjustable lenses that sit between the user's eyes and an LCD screen. This 7-inch screen is capable of reproducing

---

[92] "Oculus Rift: Step Into the Game by Oculus — Kickstarter." *Kickstarter.* N.p., n.d. Web. 14 Oct. 2013.
<http://www.kickstarter.com/projects/1523379957/oculus-rift-step-into-the-game?ref=live>.

high-resolution video and top-notch graphics, while the lenses help ensure that each eye will see a slightly different section of the screen. Combined, these features produce the 3D virtual reality experience.

Sensors present on the device detect movement of the head and allow for the in-game display to change in unison with these movements. Essentially, this allows the player to move their head in and have that movement be reflected in the game. These types of sensors, along with the 3D graphics, produce a fairly impressive VR experience that has thus far been met with an enormous amount of excitement, praise, and approval by the gaming community.[93]

---

[93] Grossman, Lev. "Hands-On Review with Oculus Rift: Virtual Reality Is Almost Here | TIME.com." *Entertainment | What's good, bad and happening, from pop culture to high culture | TIME.com*. N.p., n.d. Web. 14 Oct. 2013.

[93]<http://entertainment.time.com/2013/09/20/hands-on-with-oculus-rift-virtual-reality-is-almost-here-finally/>.

[93]"Customer Reviews: Oculus Rift Developers Kit." *Amazon.com: Online Shopping for Electronics, Apparel, Computers, Books, DVDs & more*. N.p., n.d. Web. 14 Oct. 2013. <http://www.amazon.com/Oculus-Rift-Developers-Kit-Pc/product-reviews/B00CQYZDAU>.

[93]Rasmus Kønig Sørensen. "Oculus Rift - Beyond your wildest imagination - review - FlatpanelsHD." *FlatpanelsHD - Guide to flat panel TVs & monitors*. N.p., n.d. Web. 14 Oct. 2013. <http://www.flatpanelshd.com/review.php?subaction=showfull&id=137278 2855>.

It should be noted that although Oculus certainly deserves credit for their efforts in building the Rift and bringing the device to market, the success of this device is strongly tied to the timing of its introduction. As discussed throughout this book, two large areas of recent innovation in the entertainment industry have been the rise of mobile devices and gadgets capable of producing 3D graphics. This push for mobile has led to smaller, cheaper, and more capable miniature components such as CPU's, GPU's, and LCD panels. It's no coincidence that these are exactly the types of components that go into a VR headset. Meanwhile, the movement towards 3D entertainment has similarly opened the door to more affordable solutions in that space. Putting all of this together, it becomes fairly obvious that the time is ripe for this type of technology to make the leap into the mainstream.

Oculus may be the first to market, but if recent rumors prove to be correct, it won't be long before the Rift faces serious competition from major players in the gaming industry. At present, Sony is the first industry titan to reveal that it's working on a VR headset, but it wouldn't be surprising to learn that Nintendo or

Microsoft were also hard at work on their own interpretations.[94]

Regardless of *which* company ultimately proves to be successful in this space, it's becoming increasingly clear that the VR headset is destined to become a staple in the gaming community at some point in the not-so-distant future. The technology exists and the demand from consumers is readily apparent, all that's needed now is to put all the pieces together.

Some readers may question why the Oculus Rift, a device based on mobile technologies, is mentioned here as a device that will help keep gaming consoles relevant. While it's true that VR headsets use technology that's derived from the mobile space, it's a device that simply can't be used in a mobile setting since it completely removes a user's awareness of their surroundings. Thus, even if a VR device was portable enough to wear while on the go, it would be extraordinarily challenging to use

---

[94] In March of 2014, it was announced that Facebook had purchased Oculus, further heating up the competition to be the "first to market" with a fully-functional mainstream-ready VR device.

[94]"Mark Zuckerberg - I'm excited to announce that we've... | Facebook."*Facebook*. N.p., n.d. Web. 10 Apr. 2014. <https://www.facebook.com/zuck/posts/10101319050523971>.

anyplace a person isn't completely comfortable with what was happening around them.

This sort of idea illustrates that there may not always be a fine line separating the world of mobile gaming from console gaming. Thus far, mobile gaming has been presented in this section as an entity that's completely separate from console gaming. Currently, this distinction is still apparent enough that talking about them separately makes sense. After all, there's a clear difference in quality and gameplay styles between blockbuster games found on consoles (*Halo, Call of Duty, Assassin's Creed*, etc.) and mobile hits (*Angry Birds, Draw Something, Word With Friends*, etc.). Nevertheless, it would be shortsighted to assume that this distinction will be so prominent in the future.

Mobile technologies are only going to improve. As we progress, the astounding graphical achievements and resulting gameplay experiences once reserved solely for consoles and PCs will branch out and infiltrate the mobile realm as well. As this occurs, the distinction between the two gaming "platforms" will start to diminish until it vanishes altogether.

While it could be argued that this would be the

end of "traditional" consoles, devices designed primarily with gaming in mind will almost certainly continue to be made since so many people enjoy playing high-quality games. As long as this market continues to exist, hardware manufacturers will continue making devices capable of running those high quality experiences. These devices may be mobile and equipped to handle many different tasks such as Internet browsing, phone calls, and video recording, but if these all-in-one devices are engineered to handle the latest video game releases, then it would be tough to make the case that these futuristic devices weren't a type of video game console. Furthermore, since most devices today are capable of connecting to peripherals in some capacity, it's not difficult to imagine that these all-in-one gadgets would similarly be capable of connecting to controllers, sensors, and television screens. Combining these ideas, it becomes apparent that the core principles of a gaming console aren't likely to disappear anytime soon. Of course, whether or not these all-in-one devices of the future meet the qualifications to be called "gaming consoles" is a topic that will surely be a popular area for debate between gaming enthusiasts in the years ahead.

171

Regardless of the term ultimately used to describe the devices used to play the video games of the future, it's a near certainty that there will be some extraordinary games to play on them. The reason for this is twofold:

1) Innovations like those mentioned throughout this section are likely to open the doors to entirely new experiences driven by these novel ways of interacting with games.

2) Storylines, gameplay styles, and other facets of recently –released game content have been consistently improving. Plots are thicker, characters have more depth, and overall production quality has been on the rise. If this continues, we can expect to see some truly immersive gameplay experiences.

As exciting as it may be to experience a game through the IllumiRoom device, VR headset, or otherwise, no amount of innovation can make up for a poorly conceived or executed game. Pricing, extra features, hardware add-ons, online capabilities, and the overall ecosystem provided by consoles are all-important factors that go into the decision of purchasing a gaming device, but nothing draws gamers in quite like a selection of high-quality games. In many ways this puts just as much pressure to innovate on video game developers as there is on the makers of consoles and gaming devices. To the delight of gamers everywhere, game studios have not only risen to the occasion, but developers have been raising the bar consistently nearly every year.

There's still plenty of room left for improvement and not every new feature has been well-received by the community, but recent gameplay innovations are readily apparent when comparing games of today with those produced just a few short years ago. That is, in addition to usual incremental improvements in graphics and responsiveness, there have been a number of significant changes in how developers structure, build, and create their games, as well as what gamers expect from them. The majority of these changes can be traced to one of a few larger trends in the industry. We'll take a brief look at each of these trends.

The Rise of "Blockbuster" Video Games

Over the past few decades, the video game industry, which started off as a small-time business geared towards a niche-market of hobbyists and technology enthusiasts, has taken a prominent position in mainstream pop-culture. Of course, the popularity and subsequent success of the industry has been building since the beginning. However, in recent years, this success has reached an entirely new level, which has caused many to rethink some of the long-standing perceptions about how the industry should operate. This has spurred the industry to adopt many of the same strategies and techniques employed by Hollywood when producing a major motion picture. In turn, this has resulted in the creation of what has come to be known as "blockbuster video games," or even "blockbuster franchises."

Throughout the 1990s a video game developer with a successful game on their hands could, at best, realistically only expect to sell a few million copies. For example, *Donkey Kong Country* for SNES, an extraordinarily successful game for 1994, sold a total of

9.3 million units worldwide during the remainder of the decade.[95] For comparison, *Tetris*, which was released about five years earlier (1989), sold the most copies throughout the 1990s with a grand total of roughly 26.5 million. This is certainly impressive, but it is should be emphasized that this number was achieved only after accumulating sales for 11 years.

Today's games, however, reach levels of success that weren't thought possible only a short time ago. For instance, the top selling video game of 2011 was *Call of Duty: Modern Warfare 3* (*MW3*), which sold around 22.9 million copies in 2011 alone.[96] To put this in perspective, that's only 3.6 million copies shy of Tetris' sales throughout the entire decade of the 1990s. Furthermore, assuming a $60 price tag for each copy of *MW3*, this puts total revenue around $1.4 billion in 2011.[97]

---

[95] "The 50 Best Selling Videogames of the 1990s Worldwide - VGChartz." *Video Game Charts, Game Sales, Top Sellers, Game Data - VGChartz*. N.p., n.d. Web. 21 Oct. 2013. <http://www.vgchartz.com/article/4145/the-50-best-selling-videogames-of-the-1990s-worldwide/>.

[96] "Global Yearly Video Game Chart, Week Ending 04th Jul 1975 - VGChartz." *Video Game Charts, Game Sales, Top Sellers, Game Data - VGChartz*. N.p., n.d. Web. 21 Oct. 2013. <http://www.vgchartz.com/yearly/2011/Global/>.

[97] It's likely that not all of these games were sold at the starting price of $60, but even by factoring in discounts and price variation by country, it's likely that the game still grossed over $1 billion in 2011.

Perhaps just as impressive is the fact that this amount is just under the estimated worldwide earnings of the third highest-grossing film *ever* produced: *The Avengers* (2012).[98] In other words, the earning potential of video games is now so large that it rivals the most profitable films that have ever been produced by the motion picture industry, long thought of as the juggernaut of the entertainment world. Perhaps unsurprisingly, the gaming industry is achieving this success through the use of tactics and techniques that were pioneered and refined by Hollywood.

As an example, the earliest video games tended to simply present gamers with obstacles to overcome or activities to perform with minimal storytelling involved. In *Space Invaders*, the plot was to fend off invading aliens. In *Super Mario Bros*, users were tasked with saving the princess. These plots were simplistic and really only served to provide enough context for players to understand their primary objective. A few decades later, however, and the role of storylines has been altered to

---

[98] "Marvel's The Avengers (2012) - Box Office Mojo." *Box Office Mojo*. N.p., n.d. Web. 21 Oct. 2013.
<http://www.boxofficemojo.com/movies/?id=avengers11.htm>.

contain the same type of character development, twists, and complexities that were once solely reserved for scripts written for Hollywood blockbusters.

For readers who haven't spent much time with a console lately, a few well-received games that exemplify this new breed of plot-centric gaming include *BioShock* (2007), *Assassin's Creed* (2007), *Braid* (2008), *Gone Home* (2013), and *The Last of Us* (2013), just to name a few.[99] To prevent spoiling the fun, we won't go into details about their plots. Nevertheless, curious readers are invited to research the plots of these games to see why they were included in this list.[100]

It's important to note that similarities between the film and gaming industries go far beyond their shared interest in plot-heavy entertainment. In fact, over the past few decades, one trend in the video game industry has started to blur the line between the two industries: the so-called "cinematic moment."

---

[99] "A slave obeys.." *Complex.com*. N.p., n.d. Web. 27 Oct. 2013. <http://www.complex.com/video-games/2013/09/best-video-game-plot-twists/bioshock-a-slave-obeys>.

[100] Some readers may even want to take this opportunity to do some firsthand research, which is certainly encouraged! After all, it's not every day that you get to play video games in the name of "knowledge advancement" or "research."

Cinematic moments, otherwise known as "cut-scenes," are moments in games where players have no control (or perhaps minimal control) over events occurring onscreen. These types of events are typically used to provide more enhanced visual effects than what gaming engines can handle or to advance the plot in a way that doesn't allow players to interfere. Cut scenes have been in use since gaming's earliest days, but the use of this technique has become much more frequent than what was originally found in the 1980s.[101] This inability to control the story, which makes the player more of a spectator than a participant, combined with the idea that cut scenes are becoming longer and more frequent than in the past, results in the idea that these types of moments are blurring the lines that separate the video game industry from Hollywood.

As mentioned, the use of cut scenes is somewhat controversial in that there is a rather vocal group of gamers that believe the use of cut scenes detract from the interactive experience and have become far too

---

[101] "Gaming's most important evolutions."*Page 2* -. N.p., n.d. Web. 26 Oct. 2013. <http://www.gamesradar.com/gamings-most-important-evolutions/?page=2>.

frequent in modern games. Meanwhile, those in favor of cut scenes argue that these moments add context and meaning to the interactive parts of games and can even provide some of the most exciting moments through the use of a particularly dramatic or unexpected scene. Regardless of personal tastes, however, the precedence for including cut scenes in games has been set for many years, and recent trends show no sign of a decline in the use of such moments. This will certainly be a disappointment to some gamers, but it's important to remember that changes are occurring in the video game industry on a near-constant basis. Most of these transitions are incremental in nature, but there's always the possibility that some major event will come along and shift the status quo.

Online Gaming for All

One such event was witnessed in 2004, when the wildly successful *World of Warcraft* (*WoW*) was released by *Blizzard Entertainment*, sparking our next major industry trend of discussion: massively multiplayer online role-playing games (MMORPG's).

For readers unfamiliar with the concept, an

MMORPG is an online game where players interact with others through the use of an avatar, a digital character representing the player in the virtual world. For instance, in *WoW*, each player can choose from 13 different fantasy-based character types including Dwarfs, Orcs, Gnomes, and Goblins.[102]

Gameplay largely consists of taking on quests, collecting items, and advancing the abilities of the character.[103] Players can team up with friends, socialize with other players, play mini-games, or just kill some time by exploring the massive (and visually impressive) virtual world. Some players have even managed to earn money that can be spent in the real world by finding and selling rare items in the game.

In fact, due to the staggering number of potential activities, even a player who played regularly would likely find themselves with plenty left to explore after many months of play. As a testament to this, nine years after the release of the game, there are still nearly 7.7 million

---

[102] "World of Warcraft." *World of Warcraft*. N.p., n.d. Web. 28 Oct. 2013. <http://us.battle.net/wow/en/game/race/>.

[103] "World of Warcraft." *World of Warcraft*. N.p., n.d. Web. 28 Oct. 2013. <http://us.battle.net/wow/en/game/guide/>.

gamers with active subscriptions.[104]

Now, to be clear, *WoW* wasn't the first massively multiplayer online role-playing game. That honor belongs to *Ultima Online* (1997), *Neverwinter Nights* (1991), or *Island of Kesmai* (1985) depending on how the term MMORPG is interpreted.[105] Nevertheless, *World of Warcraft* has unquestionably been the most popular, hitting 12 million active subscribers at its peak in 2010.[106]

Considering that *WoW* charges each subscriber a monthly fee ($15 in the U.S. at the time of writing), this impressive number of subscribers quickly added up to a rather sizable amount of recurring revenue for *Blizzard*. Predictably, it didn't take long for competing game companies to realize the potential for profit and adopt *WoW*'s formula when creating their own MMORPG's. Ultimately, this set the stage for a seismic shift in the industry as an onslaught of similar titles appeared, each

---

[104] "World of Warcraft Down to 7.7 Million Subscribers." *IGN*. N.p., n.d. Web. 28 Oct. 2013. <http://www.ign.com/articles/2013/07/26/world-of-warcraft-down-to-77-million-subscribers>.

[105] "Press Releases." *Blizzard Entertainment:*. N.p., n.d. Web. 17 May 2014. <http://us.blizzard.com/en-us/company/press/pressreleases.html?id=2847881>.

[106] "Gamasutra - World Of Warcraft Hits 10 Million Subscribers." *Gamasutra Article*. N.p., n.d. Web. 27 Oct. 2013. <http://www.gamasutra.com/php-bin/news_index.php?story=17062>.

hoping to cash-in on the genre's newfound success. Some of these titles earned a modest profit and garnered a respectable fan base, but none were able to eclipse the popularity of *World of Warcraft*, which has managed to firmly hold its position as a leader in the MMORPG gaming category for nearly a decade.

To be sure, *WoW's* popularity has started to decline in recent years, with many now wondering how much longer the title will retain its superior status. While there are a multitude of factors that have culminated in *WoW's* slow decline, one particular event, which stands out as an industry-changer, has undoubtedly played a major role: the release of *Minecraft* in 2009.[107]

A New Game Development Formula

Minecraft, like *WoW*, is an online-based video game which has managed to achieve unprecedented levels of success. Both games gained impressive followings, were recognized with countless industry awards, and by many accounts, earned the status of pop-culture phenomena. However, *Minecraft's* success, by

---

[107] "Minecraft." *Minecraft*. N.p., n.d. Web. 3 Nov. 2013. <https://minecraft.net/game>.

itself, is not what makes it remarkable. *Minecraft* is truly exceptional because it achieved success despite breaking just about every industry norm and unwritten rule along.

    Minecraft was created by an as-of-then unknown independent game developer working alone in Sweden. *World of Warcraft*, on the other hand, was a high-profile project crafted by countless teams of developers at one of the most respected game development companies in the world. *WoW* was advertised extensively and released with a great deal of fanfare in 2004. The team behind *Minecraft* claims that no money was spent on advertising and the only way anyone knew that the game had been released in 2009 was through grassroots initiatives and word-of-mouth marketing.[108] *WoW* requires users to pay a monthly subscription fee, while *Minecraft* offers online play for free after initial purchase of the game. *WoW* contains a vast virtual world of stunning visuals and impressive graphics scattered throughout, while *Minecraft* is known for its blocky and overly-simplistic graphics.

    In short, *Minecraft* and its unprecedented success

---

[108] "Minecraft." *Minecraft*. N.p., n.d. Web. 3 Nov. 2013. <https://minecraft.net/game>.

broke just about every industry norm that was reinforced by the largest name in the industry since 2004. *Minecraft* didn't just alter the video game industry playing field, it created an entirely new playing field and started playing on it all by itself. This game re-wrote all of the rules, achieved an unfathomable level of success, and changed the course of the entire video game industry.[109]

In terms of gameplay, *Minecraft* is an enormous game with nearly unlimited possibilities. As such, any attempt to describe the game in a paragraph will necessarily fall short. Nevertheless, the underlying concept of the game is to allow players to craft raw materials into structures.

There are two basic game modes: creative mode and survival mode. Creative mode gives players an infinite supply of materials to create any structure they can dream up. Notably, players from around the world have used this game mode to create true works of art. For example, players have managed to create replicas of entire real-world cities, landmarks, and even a partial

---

[109] Just under 12.7 million copies of Minecraft have been sold at the time of writing. For more information, see: "Statistics." *Minecraft*. N.p., n.d. Web. 10 Nov. 2013. <https://minecraft.net/stats>.

computer circuit that could perform basic binary computations.[110] As always, readers are encouraged to check out some of these masterpieces for themselves.

Survival mode, which is more similar to a traditional RPG, places players in a setting where they must build structures to survive an invading horde of zombies. Players of this mode have probably produced less imaginative structures than while playing creative mode, but this survival scenario has likely drawn in a good portion of the active players and helped Minecraft appeal to all types of gamers.

It's important to note that some readers may question the idea of comparing *WoW* with *Minecraft*. However, the purpose of comparing these games wasn't an attempt to show their similarities. The purpose was to show just how different of an "open-world" game *Minecraft* was from the industry norm that was set five years earlier by *WoW*. Furthermore, even with that break from the norm, *Minecraft* achieved a level of success that has undoubtedly changed the expectations of game developers and gamers alike, in many ways.

---

[110] "16-bit ALU in minecraft." *YouTube*. YouTube, 28 Sept. 2010. Web. 10 Nov. 2013. <http://www.youtube.com/watch?v=LGkkyKZVzug>.

Minecraft may not have been directly responsible for the decline of *WoW* in recent years, but it's a safe bet that as the revolutionary nature of *Minecraft* was revealed, it altered the expectations of gamers across the gaming community from that point forward. These changing expectations set the stage for an era of open-world games that leaves much more control in the hands of the players than what has been possible in the years leading up to the release of *Minecraft*. In addition, *Minecraft* paved the way for independent game developers to have a greater shot at success than at any other point in recent history. Indie games are trendier than ever, and as a result, investors are more willing to take a chance on independent projects, while gamers are eager to discover the next big thing from the indie scene. It's far too early to tell whether indie games will be able to hold onto their popularity in the long-term. In a more immediate time frame, however, the continued popularity of indie games is all but a sure thing.

MMORPGs and open-world games have become quite popular in the gaming community recently, but they are far from the only "trendy" gameplay concept that's been seen in the past few years. Other notable trends include post-apocalyptic games, warfare games, and first person shooters.[111] Within all of these genres however, lies another major trend that's gained an enormous amount of momentum in a relatively short period of time: downloadable content.

Downloadable Content and Micro-Transactions

Downloadable content, otherwise known as DLC, is a somewhat controversial in-game feature that most gamers seem to tolerate, but is despised by many. The essence of the feature is to allow players to buy additional content to supplement a game's core offerings. At first, the controversial nature of this feature may not be readily apparent. After all, most gamers would jump at the chance to add more levels, items, or playable characters to a game they love. However, as more and more popular games have added support for DLC, questions have come up about whether downloadable content was being purposely left out of games in order to increase profits. Worse still, some

---

[111] "The Five Biggest Gameplay Trends Heading into the Next Generation."*Fearless Gamer*. N.p., n.d. Web. 11 Nov. 2013. <http://www.fearlessgamer.com/2013/06/27/the-five-biggest-gameplay-trends-heading-into-the-next-generation/>.

gamers have speculated that content was actually being removed from the final product before release, just to squeeze a few extra dollars out of gamers.[112] In other words, questions have been raised about whether or not games would have already contained the extra content, without the need for shelling out extra cash, if DLC hadn't become commonplace in the industry.

Unfortunately, questions like this don't have easy answers. It's not possible to definitively say what would have happened if DLC hadn't become standard practice. DLC *has* become a staple in the industry and it's not likely to disappear anytime soon. Game developers love it because it gives them an extra source of revenue, which allows them to continue improving and expanding upon their games.

Some gamers are quick to demonize DLC and declare it as nothing more than extortion for content that should have been included in the main product. However, it is entirely possible that the extra content is *only* available to consumers *because* the DLC pricing mechanisms allow game companies to recoup their

---

[112] "The Great DLC Controversy." *VGUTV Video Games*. N.p., n.d. Web. 14 Nov. 2013. <http://www.vgu.tv/2013/03/06/the-great-dlc-controversy/>.

investment. That is, perhaps the only reason the developers could afford to make those extra levels, characters, or items is because they can earn enough revenue on those features to make their investment economically feasible. Clearly this is going to vary on a case-by-case basis, but it's nonetheless worth pointing out that under the right circumstances, DLC is an industry trend that greatly benefits everyone involved -- gamers included.

Regardless of whether or not you agree with the growing practice of developers earning extra revenue via DLC, one thing is for sure: the practice has been profitable.[113] As such, it appears that it would take a fairly seismic movement within the industry to cause a shift away from the use of DLC in the near future. Events like the emergence of *Minecraft*, however, demonstrate that these large shifts are certainly still possible in today's gaming industry. Nevertheless, due to the profitability of DLC, combined with the strong growth of the trend over the past decade, the likelihood of a shift away from it

---

[113] "EA: DLC revenue "above and beyond" physical sales; FIFA DLC earned $30m."*GamesIndustry International*. N.p., n.d. Web. 14 Nov. 2013. <http://www.gamesindustry.biz/articles/2010-09-15-ea-dlc-revnue-is-above-and-beyond-physical-sales-fifa-dlc-earned-USD30m>.

seems quite low in the foreseeable future.

DLC is but one of the many trends that have helped redefine the face of gaming content in recent years. The meteoric rise of open-world games, first lead by *World of Warcraft* and then more recently by *Minecraft*, has been one of the other top contributors to this reshaping of the industry. Shooters and post-apocalyptic games have been extraordinarily successful in the past few years, but it would be difficult to suggest that an abundance of gameplay innovations are what have been driving this success.

As we've seen in this section, the impact of innovation on gaming content is meaningful and helps shape the industry's direction. However, the impact of technological innovation on the *content* of video games is really quite limited when compared to some of the other topics discussed in this book. No amount of innovative gameplay mechanics, trendy features, or other technological improvements will make up for a lackluster plot or gameplay experience. With this in mind, it seems likely that in the coming years, the trends discussed throughout this chapter will certainly help mold the next generation of gaming, but the driving force behind desirable gaming content will likely continue to be what it has always been: a great storyline with an enormous amount of attention to detail, care, and precision.

Gaming as a Learning Tool

Moving away from mainstream video games, it becomes more apparent that a completely different type of innovation has been occurring over the past few years in the realm of video game development. This innovation is centered on the idea that video games are capable of fostering the development of real world skills.

A great deal of promising research has helped bolster this idea in recent years, while adding momentum to the growing trend. Some fields of study, such as driver's education and commercial airline pilot training, have been well aware of the benefits of using video game simulations for decades, but this idea has recently begun to branch out to other areas of expertise in ways that may come as a surprise.

For example, in 2012 a study was conducted which found that tenth-graders who played "two hours or more" of video games per day matched the performance of surgical residents in their ability to

191

operate a robotic-surgery simulator.[114] The fascinating aspect of this study was that these young adults had acquired their exceptional hand-eye coordination abilities by playing commercially available, run-of-the-mill video games intended for entertainment. That is, rather than building up these skills through hours of practice on a surgery simulator, these individuals gained these abilities playing games like *Call of Duty*, *Halo*, or *Gears of War*.[115]

As this study demonstrates, many of the same coordination skills built-up through countless hours of gaming may actually be quite useful in real-world activities. This particular study examined the abilities of gamers to control surgical machinery, but the operating room isn't the only place being looked at as a possible frontier for gaming skills.

Another career path that's reportedly actively seeking out and recruiting gamers is the armed forces.

---

[114] "Study: High-School Video Gamers Match Physicians at Robotic-Surgery Simulation." *Slate Magazine*. N.p., n.d. Web. 17 Nov. 2013. <http://www.slate.com/blogs/future_tense/2012/11/21/teenage_video_game_players_match_physicians_at_robotic_surgery_simulation.html>.

[115] Full disclosure: These are just examples of commercial games to illustrate the point that these individuals had been playing games commonly available to the public. The participants in the study may or may not have actually played these *particular* games.

Specifically, it's been found that gamers typically have a skillset that's particularly applicable to a career as a drone operator.

For readers unfamiliar with this term, a "drone" is the common name for unmanned aerial vehicles (UAVs). These types of craft recently made headlines in the tech world when Amazon announced an initiative to deliver merchandise to customers by using drones within the next few years.[116] However, drones are probably most known for their role in military operations, especially those of the U.S. government.[117] This usage has steadily increased, resulting in a rise in demand for UAV operators that's expected to continue well into the future.[118]

While at first it may appear that aircraft pilots would make the most natural fit for this type of role, the idea of putting an avid gamer in control begins to make

---

[116] "Amazon says drone deliveries are the future." *CNNMoney*. Cable News Network, 1 Dec. 2013. Web. 6 Apr. 2014.
<http://money.cnn.com/2013/12/01/technology/amazon-drone-delivery/index.html>.

[117] "Drone Wars UK." *Drone Wars UK*. N.p., n.d. Web. 17 Nov. 2013.
<http://dronewars.net/aboutdrone/>.

[118] "Anticipating domestic boom, colleges rev up drone piloting programs." *NBC News*. N.p., n.d. Web. 6 Apr. 2014.
<http://investigations.nbcnews.com/_news/2013/01/29/16726198-anticipating-domestic-boom-colleges-rev-up-drone-piloting-programs>.

much more sense when considering how these craft are operated.

Typically, operators sit in a room surrounded by a plethora of screens with an enormous amount of information about the drone and a handful of controls used to pilot it. Operators must stay extraordinarily focused and pay attention to every minute detail as the mission unfolds. In other words, the skillset needed to operate a drone overlaps almost perfectly with that of a gamer.

This intersection of skills has not gone unnoticed by the military. A recent report indicated trainers at an army base in Arizona expressed that individuals from the video game generation are easier to train as drone pilots.[119] Similarly, a separate news outlet recently reported that a former military commander agreed that the younger generation of video gamers are well suited for the role. According to the news outlet, the commander stated that some surveillance drone equipment is even modeled after a traditional video

---

[119] "Gamers Make The Best Drone Pilots."*Game Revolution RSS*. N.p., n.d. Web. 18 Nov. 2013. <http://www.gamerevolution.com/manifesto/gamers-make-the-best-drone-pilots-2017>.

game controller. Presumably, this design would allow gamers to have a smooth transition, while ensuring that many of their learned coordination abilities were transferred over to their new position.[120]

This career certainly won't appeal to everyone, but it is nonetheless worth noting that years of logging hours behind a console could now actually turn out to be a good investment in an individual's career. We've just touched on a few of the new opportunities being created for those with video game-related skills, but there are numerous other doors being opened every day. Eventually, video game experience may become the primary criteria by which potential applications are judged in certain fields. Amusingly enough, this would lead to job interviews where applicants would have to play video games in order to demonstrate their abilities. This may sound far-fetched, but it's a realistic possibility, especially if current trends continue.

Although we won't cover this in too much depth here, it's noteworthy that the real-life advantages gained

[120] "What's it like to pilot a drone? A lot like 'Call of Duty'." *Fox News*. FOX News Network, n.d. Web. 18 Nov. 2013. <http://www.foxnews.com/tech/2012/11/26/whats-it-like-to-pilot-drone-lot-like-call-duty/>.

by gaming aren't solely reserved for individuals searching for a career path. Over the past few years, evidence has accumulated that suggests video games are beneficial to a whole slew of things including cognitive functions, multi-tasking, the ability to focus, eye sensitivity, and even "lazy eye."[121] Of course, most of the research is still in its infancy, but the preliminary results are promising since they suggest that playing video games may indeed offer some significant benefits to gamers in the real world.

Before wrapping up this section on video games, there are a few more interesting developments worth mentioning that have the potential to gain quite a bit of momentum.

Elderly Gamers

Earlier in this section it was noted that video games have been commercially available since at least 1972, when the Magnavox Odyssey was released. This means that it's entirely possible that some folks have been playing video games regularly for over 40 years.

---

[121] "Why video games may be good for you."*BBC - Future*. N.p., n.d. Web. 23 Nov. 2013. <http://www.bbc.com/future/story/20130826-can-video-games-be-good-for-you/2>.

Individuals that fall under this category are likely at the upper end of the average gaming-age today. However, recent industry statistics indicate that the average age of a gamer is on the rise.

In 2005, it was determined that the average gamer was 24, while in 2011 that age was reported to be 32.[122] In 2013, a separate report indicates that the average age of the most frequent purchaser of games is 35.[123] As this number continues to climb, it becomes more realistic that there will eventually be a strong demand for games tailored to the elderly.

The idea of elderly people as gamers is certainly not groundbreaking, as there are undoubtedly numerous elderly individuals who play games today.[124] However, there currently aren't enough gamers in this age bracket to allow for mainstream commercial games to be tailored specifically to the tastes of that generation. As the

---

[122] "Average age of gamers continues to rise." *The Sydney Morning Herald*. N.p., n.d. Web. 24 Nov. 2013. <http://www.smh.com.au/digital-life/games/average-age-of-gamers-continues-to-rise-20111019-1m6ns.html>.

[123] "The Entertainment Software Association." - *Industry Facts*. N.p., n.d. Web. 21 Nov. 2013. <http://www.theesa.com/facts/>.
[124] Some readers may be familiar with the popularity of *Brain Age*, *Wii Fit*, and *Wii Sports* within some circles of the elderly population.

average age of the gamer continues to rise, though, it's only natural that we'll start to see more elderly individuals describing themselves as gamers. What's more, elderly people possess a number of traits that game development companies would surely find highly desirable.

First, elderly folks tend to have more discretionary income than younger individuals. Clearly, this attribute alone has the potential to motivate companies to create games specifically for the elderly. Second, since many elderly people are retired, they typically find themselves with more free time on their hands. It stands to reason that gamers with more spare time will likely purchase and play more games than those who seldom find the time to even turn on a console. These two attributes result in an extraordinary appealing demographic. The only missing ingredient is a strong demand from this age group, but that will likely change as the ages of current gamers continue to rise. As this occurs in the decades ahead, game development companies may not be able to pass up the opportunity to expand to this market.

Biometrics in Gaming

Another fascinating development related to video games is related to technologies recently used for security purposes: the field of *biometrics*.

Biometrics is a term used to describe technologies related to the collection and analysis of physical or behavioral characteristics.[125] Readers would likely be most familiar with biometric devices used to scan various body parts as a means of gaining access to some secured information or location. For example, some of the most popular biometric devices scan a person's fingerprints, voice, face, or hand.[126] That information is then processed to determine if the individual is allowed to access secured information or location. These devices are most commonly used in industries related to security, but recent reports indicate that the gaming industry may be on the verge of taking biometric technologies in an entirely different direction.

For starters, the recently released Xbox One and

---

[125] "biometrics." *Merriam-Webster*. Merriam-Webster, n.d. Web. 26 Nov. 2013. <http://www.merriam-webster.com/dictionary/biometrics>.

[126] "Biometrics." *What is biometrics?*. N.p., n.d. Web. 26 Nov. 2013. <http://www.computerhope.com/jargon/b/biometri.htm

PlayStation 4 consoles are already capable of identifying a person by "sight" using a camera or by listening to who's speaking via microphone.[127] Both of these user-identification technologies, which could be classified as a form of biometrics, open the door to a whole host of new possibilities in terms of ways for a user to interact with a console.

Even now, this domain is being explored in a few video games, such as FIFA 14, which allows gamers to communicate instructions to the rest of their team by simply speaking the appropriate command.[128] For example, it's possible to tell a team to move back or put pressure on the ball, while simultaneously remaining focused on the action unfolding in the game. It's not hard to see how this type of speech-recognition technology could just as easily be applied to other sports games or

[127] "Xbox One and PlayStation 4: Facial recognition shootout." *Reviews Games and Gear*. N.p., n.d. Web. 29 Nov. 2013. <http://reviews.cnet.com/8301-9020_7-57613371-222/xbox-one-and-playstation-4-facial-recognition-shootout/>.

[127] "PS4 Has Voice Commands Just Like Xbox One (Quick Guide)." *sidhtech*. N.p., n.d. Web. 29 Nov. 2013. <http://www.sidhtech.com/news/ps4-voice-commands-xbox-one/10025305/>.

[128] "Xbox One voice control first impression."*Product Reviews*. N.p., n.d. Web. 1 Dec. 2013. <http://www.product-reviews.net/2013/11/26/xbox-one-voice-control-first-impression/>.

any other team-oriented titles where there's a need to communicate instructions to computer-controlled characters.

The ability to communicate with games using your voice may seem like a relatively minor enhancement, but this technology is truly exciting in that it promises another element of realism. If implemented correctly, this sort of improvement could easily make an extraordinary impact on the overall gaming experience. While it's true that this technology isn't quite ready to redefine the gaming industry, it has plenty of room to grow and an enormous amount of areas in which to expand. As such, the use of this type of tech seems destined to explode in popularity over the next decade or so.

When taking a look at how biometrics may be applied to other areas of gaming, it becomes apparent that these types of technologies hold the potential to offer much more than simple voice or facial recognition.

Specifically, the area related to detecting a player's vital signs and then adjusting gameplay to match their mood seems ripe for innovation. Sure enough, in the past two years, it's been reported that at least one of the major console manufacturers has filed a patent application for gaming controllers that read information about the player holding them.[129] While it's far too early to say if or when these controllers will be used, it's not difficult to see the future foreshadowed by these types of capabilities.

If a player's heart rate was actively being monitored, games could detect which in-game actions were exciting the player the most. Gameplay could then be tailored toward those activities to provide the most thrilling experience possible. Similarly, if a game was attempting to surprise a player, it would likely be best to do so when that player is calm and not on the edge of their seat. Thus, the game could wait for the player's heart rate to reach a normal level before introducing an unexpected turn of events, thus maximizing the element of surprise.

Yet another possibility would be to give in-game characters slightly improved speed and strength while a player's heart rate is elevated to mimic the fight or flight

[129] "Sony patents biometric controller." - *Video Game News, Videos and File Downloads for PC and Console Games at Shacknews.com*. N.p., n.d. Web. 29 Nov. 2013. <http://www.shacknews.com/article/70898/sony-patents-biometric-controller>.

response that people experience while under duress. Further adjustments could also be made based on pupil dilation, the amount of sweat on palms, facial expressions, or frequency of breath. Of course, the list of ways to collect, analyze, and utilize data about players is nearly endless.

In many ways, the potential of using biometric data to enhance the gaming experience is only limited by the imaginations of game developers. The possible uses described above surely only scratch the surface of what will be seen when the technology matures. The real challenge in this area over the next few years will likely be finding a way to collect biometric data in real time, while ensuring the process is convenient and works consistently for all people.

The patent application mentioned above seems to overcome the issue of convenience by building sensors right into the controller. A player would seemingly not have to attach extra sensors to play biometrically-enhanced games. The issue of consistency, however, does seem to present some real challenges at this point in time. As a case study, let's consider Nintendo's recent failed venture into this area.

In 2009, rumors began to swirl on the Internet and within gaming circles that Nintendo was working on a revolutionary new device. That device was formally introduced to the public in June of 2009 at the Electronic Entertainment Expo (*E3*), where it was christened the "Vitality Sensor."[130] Nintendo revealed that this sensor was meant to clip onto a gamer's finger and measure their pulse. At the time, Nintendo's plans for the technology weren't particularly clear, which left many people scratching their heads and expressing confusion, rather than drumming up excitement for the device's potential.[131] If this were the only misstep by Nintendo, it's possible that the Vitality Sensor's fate may have turned out quite differently. However, in time, it also became evident that the product couldn't provide reliable results.[132]

---

[130] "Wii Vitality Sensor canned, confirms Nintendo boss."*ComputerAndVideoGamescom Multiformat RSS*. N.p., n.d. Web. 1 Dec. 2013. <http://www.computerandvideogames.com/417981/wii-vitality-sensor-canned-confirms-nintendo-boss/>.

[131] "E3: Wii Vitality Sensor revealed."*ComputerAndVideoGamescom Multiformat RSS*. N.p., n.d. Web. 1 Dec. 2013. <http://www.computerandvideogames.com/216602/e3-wii-vitality-sensor-revealed/>.

[132] "Wii Vitality Sensor canned, confirms Nintendo boss."*ComputerAndVideoGamescom Multiformat RSS*. N.p., n.d. Web. 1 Dec.

According to Nintendo employee Satoru Iwata, the device was unable to function correctly when used by approximately 10 percent of people. In the end, this flaw led Nintendo to cancel the Vitality Sensor in 2013, before it was ever released to the public.[133] Nevertheless, with two other major players in the industry presumably exploring the technology, it seems safe to say that the death of Nintendo's biometric device doesn't necessarily mean that pulse-sensing gadgets won't find their niche in the gaming community. Of course, only time will tell whether any other companies are able to *successfully* integrate this type of technology into their gaming platforms and allow the innovation to fulfill its true potential.

Gaming as Spectator Sport

One recent trend in the gaming industry that's beginning to show a lot of potential is the idea of gaming as a spectator sport. Whereas the entertainment value of video games has traditionally focused on *play*, there's a somewhat surprising, yet rather noticeable uptick in the

---

2013. <http://www.computerandvideogames.com/417981/wii-vitality-sensor-canned-confirms-nintendo-boss/>.

[133] Ibid.

demand for *watching* games played by others.

Typically, highly skilled players or particularly exciting tournaments are the most popular of these video streams, but even casual gamers will occasionally find an audience if they're playing a newly released game or provide informative commentary. A handful of websites have sprung up in an attempt to fulfill this demand, but one website in particular has managed to stand out as a leader in this budding entertainment segment: Twitch TV.[134]

Launched in 2011, Twitch has quickly grown into the largest and most popular "e-sport" viewing platform available today, welcoming more than 40 million visitors per month.[135] Notably, this volume of traffic is made even more impressive when considering that this channel of entertainment didn't even exist a few years ago.

Today, Twitch hosts live streams 24 hours a day, seven days a week, while also maintaining a library of previously broadcasted streams that are viewable anytime. In other words, Twitch offers more than enough

---

[134] "Twitch." *Twitch*. N.p., n.d. Web. 9 Dec. 2013. <http://www.twitch.tv/>.

[135] "The Official Twitch Blog   Home » The Official Twitch Blog." *The Official Twitch Blog*. N.p., n.d. Web. 8 Dec. 2013. <http://blog.twitch.tv/>.

content to keep just about any gaming fan entertained around the clock.

As always, readers are encouraged to take a moment to visit www.twitch.tv and see what kind of streams are offered. For readers who have never visited the site, the type of content found on Twitch varies quite considerably. However, the typical steam seems to contain a live feed of a game being played, alongside a real-time stream of the player and a chat window. Viewers of the stream can then communicate with the gamer as the game is played, if they wish. Some gamers provide commentary, others have conversations with spectators, and some prefer to stay quiet and focused on playing their game.

As might be expected, some of the regular broadcasters found on Twitch have become more popular than others. In fact, some have managed to gain a modest following or even earn minor celebrity status. Gamers with remarkable skills or otherwise entertaining gameplay naturally gain more viewers than others. Nevertheless, some innovative broadcasters, without any particularly exceptional gaming skills, have managed to gain an impressive following by making their streams a more interactive.

For example, one stream encountered while researching this section featured an arcade-style fighting game with a link to a website where users could place bets on the outcomes of matches. There were just under 1,500 viewers of this stream. That's well under the 32,500 folks watching the top broadcast of the night, but it nonetheless demonstrates that the platform is versatile enough to allow just about anyone to become successful with enough effort, skill, or creativity.[136]

Another interesting stream that's gained quite a bit of attention recently is known as "Twitch Plays Pokémon."[137] This broadcast showcases a live feed of Pokémon, but the game isn't being played by its broadcaster. Instead, a computer program is set up to read commands input by the stream's spectators. Every command is read in succession and then performed in the game automatically. With numerous individuals partially playing the game simultaneously, it's a curious

---

[136] The top stream of the night was a professional gamer playing League of Legends.

[137] "Twitch Plays Pokémon." *Twitch.* N.p., n.d. Web. 6 Apr. 2014. <http://www.twitch.tv/twitchplayspokemo

blend of spectating and interacting. Further, it's quite difficult to make any sort of progress in the game.

Nevertheless, it does seem to be somewhat entertaining. While researching this section, there were about 6,000 individuals watching or participating, which resulted in the game being played in an essentially random fashion. The term "social experiment" has been applied to this video stream channel, and after viewing the stream firsthand, the term does seem quite appropriate.

Attempting to determine how sites like Twitch.tv may impact the future is particularly difficult since the world of e-sports is still quite young and volatile. The impressive number of viewers and relatively entertaining content provided by sites like Twitch.tv suggest that this type of entertainment is likely here to stay, but as the constant rise and fall in popularity of video games demonstrates, gamers tend to become bored rather quickly -- regardless of how exciting or entertaining the content is initially.

On the other hand, the gaming culture is now prominent enough in modern society to support this new type of entertainment. The success of sites like Twitch make it obvious that there's now a fairly constant demand for this kind of entertainment. As a result, it seems highly likely that this market will at least maintain the popularity that it has achieved over the past few years. In fact, due to the rather sizable untapped entertainment value of this flavor of innovation, it wouldn't be the least bit surprising to see its popularity *increase*, perhaps even to the point of rivaling some traditional television programs. Regardless of how it progresses, this will surely be yet another extraordinarily exciting area of cultural and technological innovation in the decades ahead.

Total Immersion Gaming

Looking quite a bit further into the future, we arrive at the last topic of this section on video games. At present, this subject is still far more science fiction than science fact, but it's worth mentioning since it would open a whole new level of opportunity for the industry if it's ever implemented. In addition, this type of idea is particularly fascinating because it's the end goal that every video game hoping to provide a realistic experience has been moving toward for decades: total immersion of consciousness.

The idea behind this concept is that at some point, it's possible that technology will be capable of providing the brain with the appropriate signals as to make it seem -- from a player's point of view -- that they're inside and interacting with an entirely virtual world. In reality, the person's brain would be communicating with a computer, which would intercept and interpret brain signals, and send signals back to the brain as appropriate. In this way, a person would be able to interact with an entirely virtual environment in exactly the same way as they do with the real world. In other words, this type of setup would allow a person to experience a simulated environment indistinguishable from reality.[138]

Since the simulation would be entirely computer-driven, the behavior of this virtual environment would not have to mirror that of the real world. Rather, the types of environments that players interact with could be as varied as what's found in video games today. Players

---

[138]Readers familiar with *The Matrix* movies will find this type of technology familiar. However, some readers may be surprised to find that this type of tech is not quite as far-fetched as it may initially appear.

could be transported to the old west, thrust into a role as a soldier, or placed in a magical world of fantasy. The opportunities and possibilities for exciting gameplay concepts would be just as great as they are in today's video games. The main -- and most exciting -- difference would be the player's revolutionary perspective and how they could interact with the virtual environment.

Gaming while using this radical scheme would mean that taking a step or swinging a sword at the enemy would be accomplished using the same thought processes and motor skills as in the real world. If a player wanted to turn their head in the virtual world, it would be just as simple as turning their head outside of the game.

This breed of game-control would allow much more mundane games to become quite appealing. For example, games centered on nothing more than allowing players to roam around a virtual world with enhanced abilities such as super strength, the ability to fly, or mind control would easily provide hours of entertainment as the player would feel and see the effects of these abilities exactly as if they possessed them in the real world. Essentially, this sort of video game could give players superpowers, the (virtual) life of a millionaire, or anything else that game developers could dream-up.

In this virtual world, every act, movement, and activity would be practically indistinguishable from reality since the player's perception of the most basic building blocks of interaction such as sight, touch, and sound would be manipulated to replicate their behavior in the physical world. Of course, in the virtual world, laws of physics wouldn't necessarily apply, so after meeting for a cup of virtual coffee or bite to eat, a group of friends could leave the table and fly away, walk through walls, or teleport to an entirely different location. Again, the amount of unique experiences created by this type of technology is essentially limitless and truly astounding.

Of course, this kind of virtual environment is so far removed from the current level of technology available that it may almost seem like it could *never* happen. However, it's important to note that from a technical perspective, there's currently nothing that precludes this type of technology from becoming a reality someday. There will undoubtedly be numerous challenges and roadblocks -- both technological and societal -- that will slow down or perhaps even halt this

213

process. Nevertheless, it's an exciting avenue to explore, as the potential entertainment value would surely trump just about any other gaming device in history.

Whether this type of tech ever becomes a reality is still very much an uncertainty. Due to the number of technological breakthroughs necessary and the pushback that this type of reality-escaping device would certainly receive from the more conservative sections of society, it seems unlikely that such a device would be released in the next few decades. In the much longer term, it seems abundantly more realistic that if the current rate of technological progress continues, and society continues to lean toward a more progressive culture, we will eventually see such a device publically available.

Even if the aforementioned immersive virtual reality gadget fails to materialize, the future of video games still looks incredibly bright when compared to the prospective futures of many other industries. This section has discussed quite a few of the more interesting areas of development in the gaming industry, as well as where each may be heading in the years to come, but there are countless other avenues left to investigate as the current state of nearly every topic discussed is constantly changing. The future of indie games, the latest generation of consoles, and the forefront of virtual reality technologies are all rapidly advancing and are likely to have experienced major events between the time these pages were written and when they were read. As such, I would once again encourage all interested readers to take a few moments to research any and all topics of interest throughout the book.

This section has covered a great deal of material related to the latest developments, but we have merely scratched the surface. The video game industry moves quickly and there is always more to explore!

Discussion Questions:

1) How would you define the term social gaming? What makes a game social, as opposed to just multiplayer?

2) Should 3D games become the standard? Why or why not?

3) Will second screen technology ever become a standard component in the video game industry? What role will this technology play in the years ahead?

4) Would you welcome a future where video games are downloadable in the same way as today's mobile apps? What pros and cons would come with this type of distribution strategy?

5) What task, chore, or goal would you like to see gamified? Why? Bonus: describe your ideas about how you would make that task, chore, or goal into a game.

6) Are tablets, cell phones, and other electronic devices making consoles less relevant to the gaming experience? Why or why not?

7) Do you agree with the author's assessment that augmented reality gaming will catch on with the public? Why or why not?

8) Will video game consoles still be made in 10 years? 20 years? 100 years? Why or why not?

9) Is there a future for online games requiring a monthly subscription? Has Minecraft had any impact on the future prospects of those types of games? Explain.

10) Who benefits the most from downloadable content (DLC): gamers, game developers or both? Explain.

11) Do you think we will eventually see major video games marketed specifically towards the elderly? Do you think this age group will ever overtake the younger age group as the demographic that plays the most video games? Why or why not?

12) If a game was released that could detect and utilize biometric data about the player (heart rate, rate of breath, perspiration, etc), would you be more or less inclined to play that game? What do you think would be the public perception of such a technology?

13) Would society embrace a technology that allows people to experience consciousness in a virtual environment? Why or why not?

14) Would you embrace a technology that allows people to experience consciousness in a virtual environment? Would you use such a technology? Why or why not?

Chapter 7: Books, Newspapers, & Printed Media

An Industry in Transition

Printed media is by far the oldest form of entertainment discussed throughout this text. Books and newspapers have been around for centuries. Some readers may be wondering why this topic is even mentioned in a book about technological innovation. Those readers may question whether such a low-tech medium can actually be improved upon through technological innovation. The answer, of course, is a resounding "yes."

However, the improvements, and subsequently the future of printed media have almost nothing to do with the *traditional* structure of the publishing industry, which has revolved around paper. Rather, it's becoming increasingly apparent that the future of the industries dealing with printed media will almost entirely belong to digital publications. This fate is readily apparent when surveying the current state of the industry.

Over the past few years, signs pointing to an unmistakable and significant decline in the paper-based medium have been abundant. In 2011, national book retailer Borders filed for chapter 11 bankruptcy protection, and permanently shuttered its doors after years of declining sales.[139] In 2012, *Newsweek* announced that after 79 years, it would stop publishing their magazine in print.[140] Also in 2012, *Encyclopedia Britannica* announced that after 244 years, it would no longer sell printed versions of their product.[141] In 2013, the nation's largest remaining bookstore chain, *Barnes & Noble*, continued to struggle as sales declined and financial projections disappointed.[142] The long list of warning signs indicating an end to the era of printed media has continued to grow at an ever-increasing pace. Nearly every analysis, report, and measurement, coming

---

[139] "Store Closings." *Borders*. N.p., n.d. Web. 24 Dec. 2013. <http://media.bordersstores.com/pdf/Borders_Reorg_Closure_Listupdated3-17.pdf>.

[139]"Borders bookseller doubles losses." *BBC News*. BBC, 12 Oct. 2010. Web. 24 Dec. 2013. <http://www.bbc.co.uk/news/business-11966083>.

[140] Saba, Jennifer, and Peter Lauria. "After 79 years in print, Newsweek goes digital only." *Reuters*. Thomson Reuters, 18 Oct. 2012. Web. 24 Dec. 2013. <http://www.reuters.com/article/2012/10/18/us-newsweek-digital-idUSBRE89H0L020121018>.

[141] Pepitone, Julianne. "Encyclopedia Britannica to stop printing books."*CNNMoney*. Cable News Network, 13 Mar. 2012. Web. 24 Dec. 2013. <http://money.cnn.com/2012/03/13/technology/encyclopedia-britannica-books/>.

[142] "Barnes & Noble Reports Profit, but Sales Decline 8%." *NYTimes*. N.p., n.d. Web. 24 Dec. 2013. <http://www.nytimes.com/2013/11/27/business/media/barnes-noble-reports-profit-but-sales-fall.html>.

from the industry indicates that printed media is being outpaced and replaced by digital media.

At this point, it's unlikely to come as a shock to many that printed media is in the midst of what amounts to perhaps the greatest downward spiral in its history. This shouldn't be surprising considering that printed publications are more costly, bulky, and typically all-around more inconvenient than their digital counterparts.

A few books can easily weigh a couple pounds, take up an entire backpack, and tire a person out when being lugged around. In contrast, digital books are weightless and can be accessed from just about anywhere. This gives digital publications a significant advantage. It's likely that consumers have noticed and embraced these advantages, as digital book sales have steadily improved every year since 2004.[143]

Perhaps just as important for the industry is that the revolutionary nature of digital publications results in an enormous amount of room for further growth and opportunity in this budding arena. Numerous

---

[143] "Industry Statistics." *International Digital Publishing Forum*. N.p., n.d. Web. 27 Dec. 2013. <http://idpf.org/about-us/industry-statistics>.

technological innovations have cropped up to capitalize on this new generation of the industry, making it an exciting time for businesses and consumers alike. This section will explore these state-of-the-art technologies, as well as what we may see in the years ahead.

Most readers are likely well aware of many technologies that are replacing traditional printed media, since this tech is quite popular. Nevertheless, for anyone who may not be as well versed, there are a few technologies that have taken the center stage in this digital revolution.

E-books On the Rise

Tablets and devices known as e-readers are quickly becoming the "tools of the trade" for anyone who enjoys reading journals, magazines, or newspapers. These gadgets come in a wide variety of sizes and are manufactured by hundreds of companies, but they all primarily consist of some form of user-input and a screen to display the publication. If the publication happens to be a book, it's commonly referred to as an e-book.

E-books are really nothing more than a digital representation of a book. There are various formats (i.e.

PDF, EPUB, iBook, etc.), but these formats usually just dictate which devices are capable of understanding and displaying the book. From a user's standpoint, all of these formats provide a reading experience that's essentially identical.

The history of e-books is long and somewhat convoluted. Thankfully, understanding the entire history of the e-book is not vital to understanding the current status of the digital format.

It should be mentioned, however, that the earliest forerunner to the modern e-book probably appeared in the 1970s. The format started to gain popularity with some select sections of the public in the 1990s, and then really became successful commercially in the mid-to-late 2000s.[144] This brings us to today, where some very large retailers are reporting that sales of e-books are now surpassing traditional paper-based prints.[145] Globally, physical book sales are still outpacing e-books, but the

[144] "Ebook timeline." *theguardian.com*. Guardian News and Media, 3 Jan. 2002. Web. 27 Dec. 2013.
<http://www.theguardian.com/books/2002/jan/03/ebooks.technology>.

[145] "Amazon.com Says Kindle E-Book Sales Surpass Printed Books for First Time."*Bloomberg.com*. Bloomberg, n.d. Web. 27 Dec. 2013.
<http://www.bloomberg.com/news/2011-05-19/amazon-com-says-kindle-electronic-book-sales-surpass-printed-format.html>.

former's explosive growth in sales and the unmistakable downturn of the latter indicates that it's only a matter of time until e-books sit atop the sales throne.[146]

Further fueling this shift toward digital books are numerous technological innovations aiming to capitalize on new opportunities created by e-books.

For instance, one popular e-book oriented service is Amazon's "Kindle Owner's Lending Library." This initiative, pioneered by the largest e-book retailer in the United States, allows subscribers of Amazon's Prime service to borrow e-books in a way similar to traditional libraries, free of charge.[147] Users can simply choose an e-book and start reading almost instantaneously. Of course, Amazon restricts how frequently books can be checked-out and how many a subscriber can have checked out at a time, but the service still managed to draw some

[146] "E-book sales are up 43%, but that's still a 'slowdown'." *USA Today*. Gannett, n.d. Web. 27 Dec. 2013.
<http://www.usatoday.com/story/life/books/2013/05/15/e-book-sales/2159117/>.

[147] "Amazon Announces Kindle Lending Library for Prime Members." *Librarian by Day RSS*. N.p., n.d. Web. 27 Dec. 2013.
<http://librarianbyday.net/2011/11/03/amazon-announced-kindle-lending-library-for-prime-members/>.

[147] "eBook Retailers." *eBook Architects*. N.p., n.d. Web. 27 Dec. 2013.
<http://ebookarchitects.com/learn-about-ebooks/retailers/>.

concerns from critics and industry insiders over the possibility that it could have a negative impact on the income of authors and publishers.[148] Nevertheless, customers responded to the initiative in a positive manner, borrowing over one million e-books in a little over three months after the service went live.[149] Two years later, the lending library is still going strong and there's little sign of any slowdown. In fact, this innovative service has been so successful that competitors have recently started cropping-up to challenge Amazon's dominance in this somewhat unique space.

Three of the top competitors in this arena are Scribd, Oyster, and Entitle. Each offers a slightly different subscription plan and partnerships with different book publishers, but they all provide subscribers the opportunity to borrow digital books on a monthly basis

---

[148] "New Questions, Concerns About Kindle's Lending Library–What About Authors? — paidContent." *paidContent*. N.p., n.d. Web. 28 Dec. 2013. <http://paidcontent.org/2011/11/05/419-new-questions-concerns-about-kindles-lending-library-what-about-authors/>.

[149] "Over a million ebooks borrowed from Kindle Owners' Lending Library so far."*Public Libraries RSS*. N.p., n.d. Web. 28 Dec. 2013. <http://www.publiclibraries.com/blog/over-a-million-ebooks-borrowed-from-kindle-owners-lending-library-so-far/>.

without late fees.[150] While it's still too early to determine which of these, if any, will emerge victorious, it's becoming increasingly clear that e-books and the services revolving around them are turning into a highly competitive industry. If these trends continue, it's probable that coming generations of book enthusiasts will vastly prefer the convenience, payment structures, and portability of electronic copies. To future generations, books printed on paper may be thought of the same way today's youth sees vinyl records: inconvenient relics of a bygone era.

Of course, the effects of the digital revolution in printed media aren't limited to books. In recent years, there have been numerous other innovations announced that have been geared towards all types of publications.

---

[150] Greenfield, Jeremy. "Subscription Ebook Services Scribd, Oyster and Entitle Duke It Out For Early Dominance." *Forbes*. Forbes Magazine, 19 Dec. 2013. Web. 28 Dec. 2013.
<http://www.forbes.com/sites/jeremygreenfield/2013/12/19/subscription-ebook-services-scribd-oyster-and-entitle-duke-it-out-for-early-dominance/>.

Other Publications Transition To Digital

One budding service revolving around magazines is Next Issue Magazines. This service allows subscribers to access over 100 digital magazines without any restrictions.[151] In many ways, it takes the business model that Netflix employs for videos and applies it to electronic magazines. Users pay a monthly subscription fee and can read as many, or as few, magazines as they want. The magazines are completely digital and can be accessed from a wide variety of devices including phones, tablets, and PCs. This type of model is intriguing in that it appears to offer a workable blueprint for how the sagging magazine industry can proceed into the next generation.[152] The industry may never see the same level of profits it was accustomed to in its heyday, but services like Next Issue Magazines offer hope that magazines, newspapers, and other publications may be able to make the transition to the digital-centric publishing era after

---

[151] "An entire newsstand at your fingertips."*Next Issue*. N.p., n.d. Web. 28 Dec. 2013. <http://www.nextissue.com/newsstand/>.

[152] "Magazines: By the Numbers | State of the Media." *Magazines: By the Numbers | State of the Media*. N.p., n.d. Web. 28 Dec. 2013. <http://stateofthemedia.org/2012/magazines-are-hopes-for-tablets-overdone/magazines-by-the-numbers/>.

all.

It should be noted that just because digital publications are on the rise and paper-based publications are declining, it doesn't necessarily mean that demand for these types of products will cease completely any time soon. There are plenty of reasons why a person or business might prefer a printed version of a publication to a digital copy.

For example, waiting rooms are known to be stocked with magazines and newspapers for patrons to peruse. In this scenario, paper copies of magazines seem much more economical than buying and maintaining a dozen tablets for patrons to use while waiting. The benefits of paper copies become even more pronounced in this situation when the cost of replacing damaged or stolen tablets is factored into the equation.

As another example, consider how many physical copies of books exist today. Many of these books can be obtained for little or no money at all from garage sales, donation centers, or libraries. These existing paper publications aren't going away anytime soon, and there will always be a demand for good literature; especially when it's free. Digital books are great in many ways, but they require some sort of electronic device in order to be read. The cost of these devices can rarely compete with free.

These are but a few scenarios where paper publications have an advantage over their digital counterparts. However, if paper-based publications hope to survive in an age dominated by much cheaper digital alternatives, they must be made more cost effective. Of course, there are already technological innovations appearing on the market that aim to achieve precisely this goal.

A pioneer in this area is a Swedish company known as Meganews, which produces the "world's first automatic newsstand kiosk."[153]

First rolled out in 2013, the kiosk allows customers to choose from over 200 magazine titles, pay with a credit card, and then walk away with a fresh copy of their chosen magazine printed on-demand, within two minutes of payment.[154] The company behind the

---

[153] "The world's first automatic newsstand kiosk." *Meganews Magazines.* N.p., n.d. Web. 28 Dec. 2013. <http://www.meganews.com/about-meganews/the-worlds-first-automatic-newsstand-kiosk/>.

[154] "The News Stand of The Future is Here." *Ricoh.* N.p., n.d. Web. 28 Dec.

technology claims the kiosk will drastically reduce the main cost incurred by traditional magazine sales: distribution costs. That is, the costs of loading, shipping, and unloading magazines can quickly add up. By digitally transferring the latest magazines to the kiosk and printing on demand, distribution costs can be reduced significantly, thereby making these paper-based magazines competitive with e-magazines. As of writing, the kiosk is still in its testing stages and is only available in Sweden. However, if consumers react positively, it's foreseeable that the technology will begin to spread to the rest of the world over the next few years.

With all of this in mind, a few clear patterns emerge when considering where the industry may be heading. First, it's quite clear that digital media is set to overtake printed media in just about every corner of the industry. From magazines and newspapers to books, digital will dominate the future. Second, any publications that survive in their printed form will need streamlining (à la Meganews Magazines) to cut costs if there's any hope of competing with the low overhead associated

2013. <http://www.ricoh-europe.com/about-ricoh/news/2013/The_News_Stand_of_The_Future_is_Here.aspx>.

with digital goods.

The future of other areas in printed media aren't quite as clear cut, but there are a few possible developments that could have major effects on the entire industry.

Digital Textbooks

One such possibility revolves around the potential opportunities found in the area of college textbook publishing and distribution. Initially, this may not sound like a topic worthy of discussion in a book about the most *interesting* technological developments, but it's important to understand that the textbook market holds some unique characteristics that make it, arguably, the area of the publishing industry in most desperate need of technological intervention.

The first of these characteristics is the price at which textbooks are sold. Anyone who's ever purchased one of these books is likely well aware that the prices can be rather steep. In more concrete terms, according to *OnlineEducation.net*, the average price of a college textbook is $175.[155] Compare this to the estimated average cost of a hardcover, adult non-fiction book, which is $26.55, and it becomes a bit more clear that there's room for innovation aimed at bringing costs down.[156]

Second, textbooks hold a somewhat unique position in the market in that students *must* buy these books, regardless of the price. This occurs because, more often than not, university professors have assigned books as a course requirement. This often leads to a situation where students must purchase the book or put themselves at a severe disadvantage in the classroom. Clearly, these types of choices are less than ideal, and it would be a great disservice to future generations of students if we don't pursue any and all potential solutions offered by technological innovation.

One obvious solution is to phase out traditional textbooks and allow digital textbooks to become the

---

[155] "Average Cost of College Textbooks [Infographic]. Yikes.." *FlatWorld Blog RSS*. N.p., n.d. Web. 29 Dec. 2013.
<http://blog.flatworldknowledge.com/2012/09/14/average-cost-of-college-textbooks-yikes-infographic/>.

[156] "Average Book Prices, 2012." *The Library Network*. N.p., n.d. Web. 29 Dec. 2013. <http://tln.lib.mi.us/dept/technical-services/acq/files/AverageBookPrices2012.pdf>.

standard. Production and distribution costs will decrease, as fewer raw materials are needed. The finished product would be transferred digitally and at minimal expense via the Internet. Theoretically, these savings could be passed on to students, thus lowering the cost of higher education, even if only marginally.

It should be noted that students may be inclined to resist a massive adoption of e-textbooks out of fear that this would destroy the more inexpensive used textbook market. This is important since many students buy discounted used books or sell their textbooks to others in order to recoup their losses. This fear isn't entirely unfounded, as it's possible to find used textbooks that cost far less than new e-textbooks. Nevertheless, if e-book publishers were so inclined, they could *most certainly* put systems in place that would allow for a used e-textbooks market.

For example, one such system could allow digital textbooks to be transferred from one owner to another a certain number of times. The publisher would set the initial price, but owners could set subsequent prices. The price would naturally decline as the book and the information within became older, but the price would also be affected by how many more ownership transfers were possible. Under such a system, publishers could actually benefit from the used book market by taking a percentage of all used book sales or charging ownership transfer fees. Meanwhile, students would benefit from lower book costs across the board. Clearly, there are numerous details that would still need hammering out, but it's important to note that a secondhand e-textbook market analogous to today's used textbook market isn't outside the realm of possibility. All involved parties would just need the appropriate motivation to set up and participate in such a system.

Possible incentives to create such a system could include tax breaks or government funding. If nothing else, customers voting with their wallet could send a clear message to publishers that there's a strong demand for this type of technology. Of course, it's anyone's guess as to how this area will ultimately progress, but it's a safe bet that the decisions made in this area over the next few years will have a profound impact in defining how e-textbooks are bought and sold for many years to come.

The innovative potential in the textbook market goes far beyond the possibility of offering digital textbooks at reduced prices. Textbooks are somewhat unique in that they typically contain a large number of detailed diagrams, charts, and images describing extraordinarily complex topics. Even though these images are usually quite useful in helping readers understand the subject matter, e-textbooks create a great opportunity to improve upon the usefulness of these diagrams. Specifically, converting static single-frame pictures into interactive models that can be examined and manipulated from within the textbook would almost certainly benefit the reader.

For example, a student studying butterflies would likely find it helpful to rotate and zoom in on a 3D model of a butterfly, rather than looking at a static image fixed on one angle. Students could tap various parts of the digital butterfly to have further information related to the indicated feature displayed on screen. Of course, this technology could just as easily apply to nearly any other subject matter. Simply substitute the model of a butterfly for an automobile, computer, human body, microorganism, or the solar system.

This type of interactive educational technology is already in use on web sites and in software, but it has not yet been standardized for use in e-books. Nevertheless, this seems like a worthwhile addition to the textbook-based educational process as it has potential to aid students in their pursuit of knowledge. In addition, this type feature may go a long way towards justifying the significant costs of these types of books. With all of the benefits in mind, this type of technology seems to be a likely candidate for innovation in the years ahead.

Digital textbooks aren't the only section of the e-book market that could stand to benefit from embedded dynamic content. It's no surprise that people are reading less now than they were a few decades ago.[157] There are more entertainment options available today than ever before, and people seem to prefer choices that involve video, audio, or interactive elements. Along with this, there's been a noticeable decrease in the average attention span of people across the board.[158] People are

---

[157] "Study: Americans Reading A Lot Less."*CBSNews*. CBS Interactive, n.d. Web. 4 Jan. 2014. <http://www.cbsnews.com/news/study-americans-reading-a-lot-less/>.

[158] Crum, Madeleine. "Our Attention Spans Are Getting Shorter, And It's A Big Problem (NEW BOOK)." *The Huffington Post*. TheHuffingtonPost.com, 24 Oct. 2013. Web. 4 Jan. 2014. <http://www.huffingtonpost.com/2013/10/24/attention-span-book_n_4151059.html>.

quickly moving onto other tasks if their current activity fails to be interesting enough to hold their attention.[159]

Decades ago, a few images peppered into a book containing thousands of lines of text was just as entertaining as most of the other available options. Today, however, books just can't compete as a viable form of entertainment to the masses. Replacing the still images found in books with animated images is most definitely not going to cause books to overtake television or video games in popularity, but it might add just enough imagery to rekindle interest in the medium. Furthermore, turning the digital page and seeing an animated GIF depicting an event taking place in the story might go a long way towards connecting with generations that grew up seeing colorfully animated content on every web page. It may seem like a small change, but adding animated images, short video clips, or interactive content to e-books could be just the spark that the industry needs to once again make books a relevant medium in the eyes of young people.

Although the task of determining how to make

---

[159] Many readers may not even make it this far into this book. Kudos to you!

exciting or captivating digital content will certainly be a monumental undertaking facing the next generation of the publishing industry, the largest challenges faced by the industry will likely emerge as old businesses attempt to restructure around entirely new business models.

For example, a decade or two ago, self-publishing a book was almost always viewed as a back-up plan for when a book was not accepted by any mainstream book publisher. Advertising and distributing self-published works was difficult, tedious, and rarely rewarding financially. Today, however, the power of the Internet, digital media, and numerous easy-to-use self-publishing tools are making many authors take a second look at self-publishing.

Tools are improving, advertising networks are growing, and the potential financial benefits are, in many cases, greater than what's typically offered by publishing houses. In other words, the rise of the Internet and the subsequent availability of self-publishing tools has started to diminish the need for one of the primary business outfits in the printed media industry: book publishers.

Nearly everything about how print media products are produced, paid for, distributed, and consumed by the customer will continue to change as the industry shifts towards digital goods. Technological innovation will surely permeate every corner of the industry, but it's likely to be particularly focused in this area. As is always the case with large industrial shifts, these changes will surely catapult some organizations into success, while other less-agile companies fall by the wayside. Those left standing will be companies that most effectively utilize technology to create and deliver the next generation of quality publications.

This brings us to the end of our discussion on printed media. This chapter was rather short, but justifiably so as the future of this area seems considerably more predictable than many of the others in this book. The digital revolution will proliferate throughout the industry, transforming the foundation of paper products into one built on devices, networks, and digital data. This transition will open the door for hybrid mediums that blur the lines between static books and dynamic web pages. As a result of this transformation, some sections of the industry will certainly become obsolete, but the print media industry, as a whole, seems to be adapting considerably well. If the business can continue adapting

239

to and embracing new technologies as it has been over the past few years, then there's little reason for concern. Rather than assuming that the digital revolution will put an end to print media, the low costs of distribution combined with the new hybrid forms of entertainment could ultimately spark a new golden age.

Discussion Questions:

1) Is the transition from physical to digital media good for society? Consider the impact on the economy, environment, and the daily life of individuals who relied upon physical media in the past. Explain why or why not.

2) Do you agree with the author's suggestion that future generations may view physical copies of books as relics of a bygone era? Why or why not?

3) Consider the potential impacts of e-books on libraries. How will the role of libraries change in the future? Explain.

4) Will the print on-demand kiosk replace the more traditional methods of magazine distribution entirely? Why or why not?

5) Would it be worthwhile to add interactive content to e-books as described in this chapter? Do you think this type of technology would assist students in learning new educational material? Explain.

Chapter 8: Other Developments in Entertainment

Our journey through the entertainment industry is nearing an end. We've already taken a look at many areas of the industry, exploring current trends and future possibilities alike. Before concluding this book, however, there are a few additional areas that don't quite fit into any of the other categories, but nonetheless deserve recognition as being ripe for truly exciting and potentially revolutionary technological progress. First of which is an arena that some people may not even realize has opportunities for technological innovation: the world of sports.

Technology in Sports

Cutting-edge technological innovation may not be the first thing that comes to mind when discussing the future of sports, but wearable devices may soon become just as much of a staple in the industry as protective gear, jerseys, or fancy footwear.

The reason behind this is quite simple: these devices hold the potential to aid athletes with their training in ways that have not been possible. Furthermore, since athletes tend to be highly competitive by nature, they will typically seize any opportunity that

243

could give them an edge over the competition. Wearable biometric sensors that provide valuable insight into real-time athletes' performance data may turn out to be the next great advantage.

To become a serious contender, regular practices and rigorous training sessions aren't only expected, but usually required. With everyone practicing this often, there are two primary effects:

1) It's difficult for an athlete to get ahead of the competition since *everyone* is putting in long hours.

Hard work may allow an athlete to remain competitive, but it won't necessarily propel them to the forefront. This can make athletes particularly receptive to any opportunity that promises to provide even a slight edge over the competition.

2) Athletes are exposing themselves to more opportunities for injury.

The human body can only take so much abuse. There's a fine line between pushing the limits of the body to become better and pushing those limits too far, thereby introducing injury.

Believe it or not, both of these issues can potentially be alleviated through the use of wearable tech.

Using Technology to Improve Performance

Using technology to address the first issue may seem insurmountable at first, but it's actually quite simple. During a game, most sports boil down to an athlete being able to run faster, jump higher, or out-maneuver their opponent. All of these actions are a result of a complex set of movements performed in succession.

For example, throwing a football requires the athlete to grasp the ball, plant their feet, wind their dominant arm back, move that arm forward as they take a step, and then release the ball. Each of these incremental steps will have an effect on how far the ball ultimately travels. Further, each of these actions has an ideal set of movements that would result an optimal throw. By collecting precise data about each individual motion that athletes make, biometric devices offer the potential to get closer to these optimal movements than ever before.

Taking this one step further, biometric devices allow much more to be analyzed than movement. That is, an athlete's overall performance will be affected by heart rate, adrenaline levels, rate of breath, fatigue, core temperature, rate of perspiration, and nutrients in the body from recently-eaten meals. Each of these data points could be gleaned from the athlete in real-time given the right biometric sensors.

Of course, merely collecting this data isn't enough. Appropriate adjustments must be made if this data is to be useful. Similarly, only paying attention to one or two of these metrics may not be enough to impact a player's performance in any meaningful way. Taking all of this information into consideration and then making relevant adjustments, however, may result in just the edge an athlete needs to take their game to the next level. In team sports, this result would be magnified even further if each player on the team used these techniques to improve their performance. The sum total of the incremental improvements afforded by this technology could easily result in a noticeable impact on a team's overall performance.

Using Technology to Reduce Injury

In a similar vein, wearable devices may unlock data that allows athletes to adjust training techniques or in-game strategies as a means of reducing the chances of injury.

For example, data may ultimately reveal that certain types of exercises lead to higher rates of injury. In response to such a finding, athletes could alter their workout routines to focus on alternative exercises, thereby reducing the chances of injury and keeping them healthy for game time.

Given the amount of injuries suffered while playing contact sports, it may even be more useful if impact data could be collected during games and then used to help determine how players are being injured. The location of the injury on a player's body, impact force, a player's location on the field, and position of a player's limbs would all be potentially useful data points to collect and analyze. This data could then be analyzed and used to build better safety equipment, modify rules, or made available to players in an effort to help them make more informed decisions while on the field.

It should be noted that some organizations are already in the process of collecting this type of data for analysis.

According to *Inc.com*, the German TSG Hoffenheim soccer team has placed sensors in shin guards, clothing, and in soccer balls to collect data on players' strengths and weaknesses.[160] Presumably, this data could be used to help coaches make more informed gameplay decisions, such as which players to put in the game at a given time, which position each player should be playing, or how long each player can typically stay on the field without needing a break. Furthermore, since shin guard sensors could measure the various impact forces felt by players throughout the game, the likelihood of injury to a given player's foot, leg, knee, or hip could likely be calculated after enough data is collected. Again, this type of knowledge would open the door to players making adjustments to their techniques as a way of reducing the chances of injury.

A similar story is unfolding in the United States,

---

[160] "The Future of Wearable Technology by Scott Jones." *Inc.com*. N.p., n.d. Web. 25 Jan. 2014. <http://www.inc.com/scott-jones/future-of-wearable-technology.html>.

where it's been reported that the National Football League plans on embedding sensors in helmets as part of an effort to better measure the forces of impact experienced by American football players throughout a game.[161] Of course, in such an aggressive sport, there's most definitely a limit on how much can be done to reduce injuries. Nevertheless, if there are relatively simple changes that could be made to reduce the amount of injuries incurred by athletes while still preserving the enjoyment and integrity of the game, then there seems to be little reason not to make those adjustments. Analysis of in-game impact data will not be a panacea for sports-related injuries, but it does certainly present an opportunity to provide the next generation of athletes with better knowledge, safer gear, and fewer injuries than what we see on the playing field today.

With high-profile organizations such as the NFL embracing the use of "wearable tech" in sports, it seems that it may only be a matter of time until the idea is tested in all types of sports around the world. As such, if

---

[161] "New helmet sensor could better protect youth football players." *NFL.com*. N.p., n.d. Web. 25 Jan. 2014. <http://www.nfl.com/news/story/0ap2000000287866/article/new-helmet-sensor-could-better-protect-youth-football-players>.

the data proves to be anywhere near as useful as it promises, there's reason to believe that the day may come where embedded sensors are not only commonplace, but just as critical to player safety as gloves, kneepads, or mouth guards.

Negative Impacts of Technology in Sports

However, the use of wearable devices in sports doesn't come without drawbacks. As is typical with cutting-edge technology, the primary drawback of this innovation would be the added cost.

Nonetheless, major sports organizations certainly make more than enough money to cover the relatively minor expenses that would be associated with embedding sensors in athletic equipment.[162] In addition, there's no reason why smaller organizations would *have* to start using such technologies in their organizations. High schools and little leagues *could* certainly continue

---

[162] "NFL: Is the Most Popular Sports League in the USA Really Too Big to Fail?."*Bleacher Report*. N.p., n.d. Web. 12 Apr. 2014.
<http://bleacherreport.com/articles/1707663-nfl-is-the-most-popular-sports-league-in-the-usa-really-too-big-to-fail>.

[162]"How The National Football League Can Reach $25 Billion In Annual Revenues."*Forbes*. N.p., n.d. Web. 12 Apr. 2014.
<http://www.forbes.com/sites/monteburke/2013/08/17/how-the-national-football-league-can-reach-25-billion-in-annual-revenues/>.

operating as they do currently. However, the real issue arises when considering the impact of these devices would only be useful to athletes who could afford to purchase such tech. Wealthier athletes, school districts, or little leagues could harness the power of biometric data sensors to improve their techniques and increase their skills, while less affluent groups or individuals would effectively be excluded from such benefits.

To be sure, a financial gap has practically always existed in the world of sports. It should come as no surprise that some athletes can afford better equipment than others. Some athletes can afford to go to professional training camps, while others must make the best of their backyard and some old free-weights bought at a discount store. Spending more money doesn't necessarily result in a better athlete, but -- everything else being equal -- money certainly can provide an individual with an edge over the competition.

As digital sensors are used more frequently as a tool in understanding an athlete's performance metrics, it stands to reason that those who could afford such technologies would disproportionately benefit. Clearly, the potential disadvantage for less affluent athletes is a concern that needs to be kept in mind as this type of technology is developed. With any luck, the devices and software can be made affordable enough to resolve this issue before it has the chance to make any significant negative impact.

Even with this potential drawback, the positive aspects of increased technology in sports could still easily outweigh the negatives. Thus far we have only considered the usefulness of sensors as they pertain to recording data about individual athletes, but the potential benefits of embedded sensors aren't solely limited to transmitting information about players. In many ways, the possibilities presented by devices, sensors, and data becomes much more intriguing when all facets of a game are considered potential targets for such technological innovation.

Using Technology to Improve the Game

Although this idea can be applied to just about any sport, let's briefly consider the implications of such an idea on American football.

A common issue faced by referees in football is the task of determining where the ball should be spotted (placed) after a play is finished. Typically, the referee will make his best guess, which will usually be accepted as being "accurate enough." The game will then continue and the next play will occur. However, on occasions where the ball is close to the first down marker or the end zone, its location can be absolutely critical to the game's outcome. In situations like this, precision and accuracy are essential in ensuring the correct decision is made.

However, as most football fans will likely attest, a person's ability to determine the position of a ball at a given moment is far from perfect. From time to time, referees will make mistakes when spotting the ball. The NFL has recognized this and attempted to correct for human error by allowing plays to be reviewed in slow motion. However, with a finite number of cameras and 22 bodies that can get in the way, the task of determining where the ball was at the end of a play can still be quite challenging.

With the idea of embedded sensors in mind, one potential solution to this problem would be to place a sensor in balls that would report its precise location at any given moment. Sensors could also be placed in the

253

first down markers, goal line, and along sidelines to allow the ball to detect whether it's crossed any of these boundaries. The referees could then simply reference this data when spotting the ball to reduce the margin of error and provide a fairer game for all players.

Taking this idea to the extreme, a sensor could be added to just about everything in the game. Each player could have sensors in their helmet, shoulder pads, elbow wraps, gloves, knee pads, and shoes. Whistles could transmit data about the exact time they were blown to determine when plays start and end. The ball, sidelines, first down markers, goal line, and uprights could all report the precise location of the ball as it relates to these other key boundaries. All of this data combined would allow referees to determine the relative position of all players and game components at any given point in time. If an adequate up-front investment of time, energy, and effort was made in this area, it seems reasonable that this type of tech could easily reduce or perhaps even eliminate the amount of human errors made while officiating sporting events.

Even if less dramatic changes were desired, adding a small handful of sensors to critical game components could go a long way towards making the officiating of football, or any other game to which this idea was applied, far more precise. Of course, introducing new technologies in the name of extra precision may not necessarily be what fans want to see in their favorite sport.

Some fans greatly value the traditions associated with their sport of choice. These individuals would likely be adamantly opposed to the idea of incorporating more technology into the game since it would require a clear break with tradition. People sharing this point of view may point out that the human element that comes with people officiating sporting events has been a part of the sporting world forever, and is as much a part of the game as the players. To these fans, improving the accuracy and precision of the officiating is secondary to the goal of preserving the traditions of the sport they love. As a result, there would almost certainly be an extraordinarily vocal group of sports fans firmly opposing the use of technology to improve officiating.

Regardless of viewpoint, it does seem that the very nature of sports creates a limit to how much of an improvement technology would be able to provide. After all, two of the main goals of sports are:

1) Test the physical and mental abilities of participants

2) Entertain fans

Introducing more technology may allow for fairer competition, but a balance must be struck between these (sometimes competing) goals so as to not diminish excitement and drive fans away. While logic seems to dictate that we should use any tools at our disposal to provide the most accurate officiating possible, some sports fans would readily sacrifice accuracy to preserve custom. These fans simply accept that referees occasionally make absolutely terrible, game-deciding calls.[163] These fans would likely argue that atrocious calls have always been part of the game and are far less destructive to the spirit of the game than the proposed sensor-enhanced officiating changes.

Ultimately, there's no "correct" answer here. Some fans will embrace the opportunity for more equal rule enforcement, while others will prefer to preserve tradition. As a result, this issue has the potential to result

---

[163] "The List: Worst calls in sports history." *ESPN.com*. N.p., n.d. Web. 8 Feb. 2014. <http://espn.go.com/page2/s/list/worstcalls/010730.html>.

in some interesting and perhaps even heated debates within sports communities in the years ahead.

The Use of Technology During Games

Another interesting debate that's just starting to gain some momentum is the idea of letting coaches use tablets, laptops, or other electronic devices during a game. While, at first, this may not seem like an issue capable of sparking too much contention, the potential implications of such a development are actually quite far-reaching.

For starters, most professional leagues currently ban the use of such equipment, including the National Football League, National Hockey League, and Major League Baseball.[164] Thus, the same social forces responsible for creating a resistance to the placement of sensors in sports equipment would likely result in a stern resistance to this change for the very same reasons. On a

---

[164] "Players, coaches told iPads are banned on game day." *ProFootballTalk*. N.p., n.d. Web. 13 Feb. 2014. <http://profootballtalk.nbcsports.com/2011/09/11/players-coaches-told-ipads-are-banned-on-game-day/>.

[164] "Sports & Social Media." : *Phoenix Suns Players Get Android Powered Playbooks*. N.p., n.d. Web. 13 Feb. 2014. <http://emilydroessler.blogspot.com/2012/02/phoenix-suns-players-get-anroid-powered.html>.

more substantive level, however, there are a few real potential repercussions to the sports world if this change were to be made.

First, if the use of mobile devices were allowed during games in an unrestricted manner, then it would open the door for coaches to tap into any and all resources at their disposal to help them make better game-time decisions. Coaches could chat with anyone in the world at any given moment during the game, research how other coaches have handled similar situations in the past, or take instant-polls from the crowd or teams of experts to make tough decisions. This added ability to communicate with the outside world may not be devastating to the structure of the game, but it could possibly diminish the role of the coach and supporting staff since the coach could receive strategies from any number of external resources throughout a game.

Even if mobile devices were restricted so that they couldn't connect to the Internet, coaches could use them to watch instant replays of the action, which could help them decide their next move. This ability would likely be found particularly useful in American football, where the coaches can "challenge" the ruling on the field if they believe officials have made an error.

Today, coaches must rely upon their staff, who watch replays, to know if a call should be challenged. However, if all coaches had tablets, they'd be able to examine plays numerous times, from different angles, and in slow-motion before deciding whether to challenge calls. Again, this wouldn't have a major impact, but it would certainly have *some* impact on the game.

Finally, if Internet connectivity and the ability to watch video were both disallowed by coaches while using tablets on the sidelines, the impact on the game could still potentially be quite large since mobile devices are very effective at crunching data, calculating probability, and using that information to formulate strategies. For example, the world of competitive chess, in particular, demonstrates that given the right software, computers can strategize extraordinarily well. With this in mind, it's not a stretch to imagine the day where coaches could pay for software to analyze the game and then recommend the optimum strategy for any given situation in real-time.

Historical game data, game strategy theory, information about the opposing team, and the stats of each individual player could all be harnessed by the software to calculate optimal strategies, which plays to run, and when to run those plays. With well-designed software, fast processors, and the right data set, this method of play calling could eventually outperform traditional coaches, who simply can't process data as fast as a machine and are also vulnerable to human error. Given the cutthroat nature of competitive sports, it wouldn't be long before this coaching strategy was adopted as standard practice if it were found to be more effective than traditional techniques. In other words, the simple act of allowing coaches to use tablets on the sidelines could potentially lead to a transformation in their role from "strategy maker" to "relay man."

Perhaps worse still, the team with the best analysis and strategy-making algorithms would hold a clear advantage over others. Since software doesn't typically come cheap, this could create a situation where teams with the most money would get the best software, while financially-constrained teams would simply be left at a disadvantage.

To be sure, there are numerous events that would need to take place for such a future to become a reality.

Given the sports community's history of resistance to major changes, it seems highly unlikely that all of these events could occur without decision makers in the sports leagues stepping in to block such actions.

A much more likely scenario seems to be that the use of tablets on the sidelines will eventually be allowed, but that usage will be heavily restricted in an effort to prevent such impactful changes to the game.

Such an approach appears to be the preferred strategy for the National Basketball Association (NBA), which doesn't bar coaches or players from using tablets on the sidelines during games. As a result, players can frequently be seen reviewing strategies or watching film on tablets from the bench, while the game continues to be played out on the court. These players will typically watch footage recorded earlier in that very game, in an effort to better understand the opposition and look for ways to capitalize on any areas of weakness.

Although this type of strategy hasn't been widely adopted yet, it's been put into practice by a number of players on the Portland Trail Blazers, where use of the technology has been encouraged across the team.

A number of players who have chosen to use

tablets during games have indicated that they've been instrumental in analyzing each component of their performance from earlier in a game and making appropriate corrections when they return to the court moments later.[165] Foreseeably, this type of strategy could one day give athletes who are particularly good at reacting to information gleaned from film an edge that just wasn't possible in the past.

It should be noted that even in the relatively relaxed NBA, there are some safeguards in place to protect the game's integrity.

For example, neither coaches nor players are allowed to watch live footage as it's unfolding. This suggests that although the NBA rule-makers are more lenient than their counterparts in the other professional leagues, they are keeping a watchful eye on how such technology is being used, as well as how it affects games. This also seems to suggest that radical developments like the automated coaching scenario previously outlined would not go unnoticed by the league. This further

---

[165] "Daily Buzz: Secret to Blazers' success just might be cunning use of tablets." FOX Sports on MSN. N.p., n.d. Web. 15 Feb. 2014. <http://msn.foxsports.com/nba/story/daily-buzz-the-secret-to-blazers-success-just-might-be-cunning-use-of-tablets-112713>.

suggests that rule makers would also ban any technology more disruptive than the tech allowing players to watch a live video feed of a game.

An Industry Resistant to Change

We have spent a great deal of time in this section focusing on just a few of the possible ways technology could impact the sports communities, but in truth, the potential impact of innovation on the world of athletics is quite large. Furthermore, as a few of the mentioned hypothetical scenarios demonstrate, even seemingly simple changes such as allowing coaches to use tablets can have a noticeable effect on how a sport is played.

Most major leagues seem to recognize this, and as a result, have been slow to adopt new technologies. If this cautious approach continues, the role of new tech in athletics could be minimized quite considerably, which would surely please a highly vocal section of today's sports fan base.

Readers who find themselves hoping to see minimal changes to their favorite sports can take solace in statements like those recently made by NFL spokesman Brian McCarthy, who asserted that the league is "strategically holding back" from using new technologies. McCarthy continued, "We try to keep as much of the human element in the game [as possible]."[166]

Similarly, according to a recent article by the Wall Street Journal, a spokesperson from Major League Baseball indicated that the league doesn't anticipate changing the rules surrounding the use of new technology, which currently prohibits players and staff from using electronic devices after the start of batting practice.[167]

With such decisive statements coming from major sports organizations, it seems the overriding opinion of the sports community is clear: the adoption rate of new technology will be slow. Unless privately-held opinions on the issue differ greatly from these types of public statements, it seems extraordinarily unlikely that we'll see an uptick in the rate of technological innovations

---

[166] Couwels, John. "Tablet computers are a game-changer in professional sports."*CNN*. Cable News Network, 12 Nov. 2011. Web. 13 Feb. 2014. <http://www.cnn.com/2011/11/12/tech/tablet-computers-and-sports/>.

[167] "Pick Up the Phone, NFL: The Future Is Calling." *The Wall Street Journal*. Dow Jones & Company, n.d. Web. 16 Feb. 2014. <http://online.wsj.com/news/articles/SB10001424052970203893404577100 683039518086>.

introduced into these major organizations anytime soon.

To a certain extent, it shouldn't come as much of a surprise that sports organizations are hesitant to adopt new technologies. After all, despite the excitement and positive impacts that typically come with new technologies, there's always a certain amount of inherent risk present when introducing any type of meaningful change. In this case, the major risk is a reduction in the number of fans, which ultimately translates into lost revenue.

Major sports organizations seem to be aware of this and have tailored their technology-adoption strategy accordingly. The guiding principle in this area appears to be fully understanding the impact of new technologies before introducing them on a large scale. While this strategy essentially guarantees that the leagues will always lag slightly behind the curve in terms of technology, it also helps ensure a smooth transition to the future.

Fans reluctant to change would likely cause less strife over smaller, incremental adjustments rather than massive changes introduced all at once. At the same time, the occasional introduction of a new technology

265

will help safeguard against falling too far behind the curve. In theory, this strategy should allow major sports organizations to continue changing with the times, while minimizing the risk of driving away large sections of their fan base.

An Industry Ripe for Disruption

However, this wait-and-see approach does come with a significant tradeoff. Since the adoption rate of new technologies in these larger leagues is expected to be intentionally slow, they'll be opening the door to other leagues that choose to incorporate cutting-edge technologies. For this reason, the sports industry seems particularly ripe for "disruption."

For readers unfamiliar with this use of the term, a "disruptive" company is one that upsets the status quo of a well-established industry. "Aereo," a company mentioned earlier in this book is a prime example, since it could shake up the profit distribution of other businesses in the industry if it were to grow in popularity. A list of the top disruptive companies is published yearly by MIT

Technology Review and can be found in the footnotes.[168]

The fact that major sports organizations are reluctant to adopt new technologies makes this industry especially prone to disruption, since any new technological breakthroughs will go unutilized. Meanwhile, a pioneering outfit can pick and choose from these innovations to create a more exciting or engaging sporting event than what's offered by the major organizations. In other words, by resisting the adoption of new technologies in their leagues, major sports organizations are opening the door for new competition.

For example, if a football league were to embed sensors all over the field, on players, and in all major points of interest, they could then advantageously market themselves as a more fair and technologically-savvy league. By continuing to adopt new tech trends as they develop, this start-up organization might be able to work their way up to a sizeable fan base by exploiting the slow adoption rate of today's major leagues.

Notably, this type of league wouldn't even have to have the "greatest" athletes, since the major draw would

---

[168] "50 Disruptive Companies 2013." *MIT Technology Review*. N.p., n.d. Web. 13 Apr. 2014. <http://www2.technologyreview.com/tr50/2013/>.

be the revolutionary technological additions to the game. Athletes would of course need to be somewhat skilled in their sport to keep fans interested, but the main entertainment value would come from novel technology-oriented additions:

- While watching a game, fans could use their handheld devices to vote on which coaching decisions should be made in real time, thereby introducing a social component to the world of sports unlike anything seen today.
- A system could be put in place to allow fans to view players' live biometric stats. This would allow spectators to see how fast their favorite athletes are moving, how hard they're kicking the ball, or how high they're jumping at any given moment.
- Some spectators might even just get satisfaction in knowing that the games are being officiated with as little human error as possible. As discussed previously, the introduction of sensors around the field and into game objects would go a long way towards realizing this goal.

These are clearly just a few possibilities. The potential additions here would only increase as technology continues to evolve.

Whether there's a strong demand for such a league is a question that can only truly be answered in time. Nevertheless, the idea of incorporating new technologies into traditional sporting events seems intriguing enough that it just may turn out to be a worthwhile target for innovation. Regardless of where this innovation comes from, if there's enough demand for increased technological involvement in sports, that demand will eventually be met. With this in mind, anyone who's both a sports fan and a technology enthusiast may want to keep an eye on this area over the next few years, as it may reveal some particularly exciting developments. As always, I would encourage interested readers to take a few moments to research the latest progress in this area online, as new technological advancements occur regularly.

This brings us to the end of our discussion on technology's role in the world of sports. It's important to note that we've only just begun to explore the technological developments that may lay ahead in this area. There certainly are numerous other ways technology may help shape the future of athletics, but the information in this section should at least suffice as a starting point for further investigation into other interesting developments.

269

Biometric sensors, other devices, and social media will be a few of the major players in this area in the coming years. How these technologies will be combined, used, and accepted by the sports community will help define the direction of technology in sports. The tendency to preserve tradition in the sports community may cause technological innovation in the sports and athletics communities to lag far behind other areas of the entertainment industry. Of course, the major players in the industry will have a sizable impact on the extent of this lag, as they will decide which technologies to adopt, when they should be adopted, and how they'll be integrated. Nevertheless, sports fans fund the industry and therefore will have the largest impact. Which technologies fans choose to embrace will ultimately decide the fate of technological involvement in the world of sports.

Total Immersion Entertainment

Switching gears completely, the final topic worthy of discussion in this chapter is an area that was touched on briefly towards the end of the discussion surrounding video games. This is an exciting area that, while certainly applicable to the future of video games, has a much broader potential impact. This topic is "total immersion entertainment."

In the context of video games, we've already discussed the basic principles of this idea, as well as a few of the ways it could affect society. As a refresher, the underlying concept here is that it may one day be possible to enter a virtual world that is nearly indistinguishable from reality. The objects in this realm would look, feel, smell, and even taste as real as anything experienced in the physical world. All of this would be accomplished by manipulating signals in the brain, which is ultimately what allows a person to determine what's happening in the world around them.

A depiction of this concept can be seen in the popular *Matrix* movie trilogy, although many of the details surrounding how this idea is implemented in those movies are somewhat far-fetched. Nevertheless, readers who have seen any of these movies will already have a good grasp on this concept.

For readers who are still struggling with this concept, take a moment to consider how you interact with the world. When you see something, light enters your eye, this information is then transmitted to your brain, which processes this data and turns it into something meaningful. A similar sequence of events occurs with all other senses. When you touch something, taste something, or smell something, your sensory organ acquires information and translates that information into an electrical signal. Ultimately, this signal is transmitted to, and processed in, the brain.

In this regard, the signals that occur in the brain during the processing of this information have the "final say" on what a person perceives to be happening around them. Fully understanding the details of how these signals operate could one day open the door to intercepting and manipulating these signals as a means of changing a person's perception of reality.

Now that we have a basic understanding of the concept, let's focus on the question of whether or not we will ever actually be able to manipulate the brain in such a fashion.

Clearly, this is the type of question that can only truly be answered in time. There are, however, a number of recent developments in this area that seem to suggest the idea may not be as far-fetched as it may first appear.

In August 2013, for example, an experiment was conducted by the University of Washington, which demonstrated that a person's limbs could be moved by electronically transmitting signals to their brain. Specifically, this experiment allowed one researcher to move another researcher's finger by simply thinking about the act.[169] For the purposes of our discussion, this study is noteworthy as it shows that the technology to directly manipulate brain signals already exists, albeit in a rather rudimentary form. In this particular study, that technology was being used to engage the part of the brain responsible for movement, but in the future it could be modified to target the areas of the mind responsible for taste, smell, or any of the other senses.

As another anecdotal example of what we're already capable of in this area, a recent report indicated that in 2012, a woman with quadriplegia was able to move a robotic arm using only her thoughts. Electrodes were implanted in her brain at the precise location where arm movement is processed. These electrodes would

---

[169] Vergano, Dan. "Researcher remotely controls colleague's body with brain." *USA Today*. Gannett, 28 Aug. 2013. Web. 2 Mar. 2014. <http://www.usatoday.com/story/tech/sciencefair/2013/08/27/human-brain-remote/2709143/>.

detect activity in that area of the brain, process that data, and then act upon it appropriately. After 13 weeks of training, the woman was able to move the prosthetic limb with enough precision to grasp and move items with a success rate of about 90 percent.[170]

Without a doubt, it will take some time for this technology to mature, but even in its current state, this demonstrates that it's possible to intercept specific brain signals, send those signals out to a device, and then act on them. In the context of our "virtual reality" discussion, this is important since we would eventually require the ability to intercept these signals and send them to the device responsible for creating the virtual environment. That is, if the person wanted to jump, all of those signals created by that action must be relayed to the machine, which would then process that jumping motion and send back the appropriate sensory information (e.g. wind on the face, sense of motion, etc.).

A third example of progress in this area comes from a 2013 *MIT News* article titled "Neuroscientists

---

[170] Choi, Charles. "Quadriplegic Woman Moves Robot Arm With Her Mind."*LiveScience*. TechMedia Network, 17 Dec. 2012. Web. 2 Mar. 2014. <http://www.livescience.com/25600-quadriplegic-mind-controlled-prosthetic.html>.

plant false memories in the brain."[171] The article describes a recent study where researchers were able to successfully implant synthetic memories in the brains of mice by manipulating specific brain cells. The full details of the study are available at the website provided in the footnotes, but the primary importance of this study is that it shows that we can detect and interact with brain signals other than those responsible for movement. Understanding all types of brain signals is crucial if we ever hope to fully interact with the brain in a non-trivial fashion.

There's surely a wealth of the other research available detailing the great strides being made in this area. Nonetheless, these few studies go a long way towards demonstrating that the idea of manipulating brain signals to create true virtual reality isn't as implausible as it may first appear. In fact, when considering the state of the research discussed here, it starts to not only seem possible, but highly likely if given a long-enough time frame. With that in mind, the

---

[171] "Neuroscientists plant false memories in the brain." *MIT's News Office.* N.p., n.d. Web. 2 Mar. 2014. <http://web.mit.edu/newsoffice/2013/neuroscientists-plant-false-memories-in-the-brain-0725.html>.

question shifts from whether this technology possible to if the scientific community, governmental bodies, and society in general would ever allow such a technology to be used.

To be sure, there are a number of reasons why people wouldn't want this technology to be explored or made available.

For starters, the very notion of a technology capable of manipulating a person's thoughts, feelings, or actions can be quite alarming, especially when considering how this type of technology could be abused. Unquestionably, this type of tech could open the door to malicious use. It should be noted that there might be limitations to how the brain could be manipulated. There may also be safeguards that could be put in place in order to prevent against some types of abusive or malicious behaviors. Nevertheless, this does appear to be a legitimate concern that needs to be kept in mind as progress is made in this area.

Another reason people may be opposed to this type of technology is the potential for it to encourage users to escape into a virtual world, rather than participating in the real world.

The motivating factor for such a decision would be that the virtual world would allow the person to do just about anything they wanted with little consequences. They could steal cars, rob banks, or perform nearly any other action they desired without repercussions, since all of these activities would be taking place inside of a virtual environment running on a computer. Conceptually, this wouldn't be any different than running around and causing mischief in virtual worlds created by today's video games. The distinguishing factor, however, is that this virtual world experience would be nearly identical to the physical world.

After experiencing a world with no consequences, some people may simply decide that the experiences afforded by the real world just can't compete. If enough people were to opt-out of the physical world, it could potentially have a ripple effect on numerous other areas of society, including decreased productivity, a large spike in antisocial behavior, and a potential increase in psychological disorders resulting from prolonged exposure to a world without consequences. Again, these are merely *potential* concerns and are by no means

277

guaranteed. Nevertheless, these all seem like areas worth monitoring as the technology advances.

Finally, some people may be against this type of technology for the simple reason that they believe very little good can come from it. Such folks may argue that even if all of the other concerns fail to materialize, this type of technology should not be pursued since it has no measurable positive impact. Further, the technology would cost time, energy, and resources that could otherwise be spent advancing more worthwhile endeavors.

While it's ultimately up to the reader to decide which side of this debate is most agreeable, it should be mentioned that there are some potential upsides. For example, in addition to the initial economic benefits that this technology would create as a result of companies creating hardware, software, and accessories, it could provide a fantastic platform for building and mastering skills for jobs that come with a large amount of repercussions in the real world.

For example, some professions -- think airline pilot, surgeon, or soldier -- require many hours of training to hone skillsets. In such occupations, one mistake can be the difference between life and death. As such, it's desirable that individuals entering these fields gain as much experience as possible before putting their skills into practice. Building these abilities in a virtual world would have some clear advantages since the environment would be nearly indistinguishable from the real world, but the consequences for mistakes would be greatly reduced. These characteristics will likely make the thought of training in a virtual environment highly appealing to anyone interested in improving their skills in a low-risk environment.

Another area where this type of innovation could potentially have enormous benefits is a subset of the elderly or disabled communities.

This notion is derived from the fact that some of these individuals would be able to navigate within the virtual world as though they had their full range of mobility restored. That is, as long as person's brain was sending the appropriate signals, that person could walk, talk, move, jump, speak, or otherwise maneuver as they pleased in the virtual world -- even if their physical body was incapable of performing such actions. This would be possible since the brain would be the only driving force

279

behind the actions that a person would take. The body need not be capable; only a functioning brain would be required.

Combine this technology with the idea of multiple people being able to "log-in" to the same virtual environment, and it's foreseeable that this type of tech could one day allow a father to take a walk with his wheelchair-bound child, or a bed-ridden elderly couple to take a drive down the countryside within a virtual world. Hence, there does appear to be at least *a few* potential benefits that could arise from such an innovation's mainstream support.

Whether these benefits outweigh the negative aspects of this technology, however, is ultimately up to the reader. The likelihood of such an innovation actually being developed may still be uncertain, but there's a real possibility that it, or something similar, will become a reality. As we move closer towards this type of capability, it becomes increasingly important to be aware of the potential impacts on society, as they could be severe. A current understanding of these issues provides the longest possible time to determine their implications and perhaps even influence the direction of such technologies for the better.

With enough care, effort, and research, well-informed individuals could minimize the negative aspects of this, or any futuristic development, while maximizing the positive impact on society. Readers of this book have already taken their first step towards understanding these issues and may find themselves to be particularly well equipped to provide such guidance as these developments unfold.

Discussion Questions:

1) Do you agree with the author's assertion that biometric devices could give wealthier individuals an unfair advantage in sports? If so, should an effort be made to "level the playing field," or is this type of advantage inherent in sports? Explain.

2) Should sensors and other devices be used to improve the accuracy of officiating during sporting events? What advantages and disadvantages of this innovation do you view as being the most important? Explain.

3) Should the use of tablets and other mobile devices by athletes and coaches on the sidelines be prohibited during the game? Why or why not?

4) Will the major sports leagues ever allow technology to play a larger role? Why or why not?

5) Do you think fans would embrace a greater presence of technology in sports? Explain.

6) Do you agree with the author's suggestion that we may see smaller leagues appear with an emphasis on integrating more technology into the games?

7) Will society ever embrace the use of technology to create true

virtual reality?

If so, how do you see the use of such a technology being restricted? If you don't see the use of such technology being restricted, explain the advantages and disadvantages of allowing unrestricted access to virtual reality environments.

Chapter 9: Conclusion

We've reached the end of our journey through the world of technological innovation in the entertainment industry. What a journey it's been!

We launched our exploration of this area by taking a look at some of the largest recent trends in the industry. For example, the transition from physical to digital media, combined with the rise of the Internet, has resulted in an industry-wide shift in the way media is produced, distributed, and monetized. Similarly, the push towards social-based entertainment has been reaching a fever pitch, with just about every form of entertainment attempting to capitalize on this fashionable industry trend.

Next, we visited the television industry and saw that one of the largest developments in this area has been the shift towards Internet-based TV. Despite the current limitations and drawbacks of Internet television, the ever-growing "cord cutter" movement has been gaining quite a bit of momentum. With the current leaders in this area now becoming household names (i.e. Netflix, Hulu, Amazon Instant Video, etc.), traditional television industry leaders have started to release their own online streaming services in an attempt to prevent further loss of customers. Whether these efforts prove successful is, at this point, still an uncertainty. Nevertheless, these developments all but guarantee that this will continue to be a particularly active area for innovation over the next few years.

In the realm of music, we delved into one of the most complex industry-wide challenges that arose in the past decade: online music piracy. As in the television industry, the shift towards digital media has had an extraordinary impact on how the music industry distributes and monetizes content. This shift has created some unique opportunities for content creators, but has also opened the door for large-scale media piracy operations. How both of these areas progress will undoubtedly shape the future of the industry.

We also saw that the technologies used to create music have continued to evolve over the past decade, particularly in the areas related to digital music creation. This has paved the way for new and exciting music production techniques such as the digital guitar and the music glove.[172] These developments have set the stage

---

[172] The specific digital guitar discussed in this book is the now-deprecated

for digital-based music to flourish in the coming years.

A tour of the hottest trends in video games revealed the industry is bursting with excitement. There has been a rather large shift towards mobile gaming that has had a noticeable impact on the popularity of traditional portable gaming consoles. Meanwhile, the impact of "social" on the gaming world has steadily grown, with online multiplayer games continuing to be a massive success.

The role of the gaming console has been continuously transitioning away from being a "gaming" device towards an all-in-one "entertainment" gadget. New features in the latest generation gaming consoles, such as the ability to watch cable TV on the Xbox One, have only solidified this transition.

We then saw how video games and video gamers are being recognized for real-world skills acquired as a result of gaming. The potential applications of these abilities are only now starting to be realized, but it's expected that the advantages offered by the skillset of gamers will be further identified and capitalized on in the

---

Misa Kitara. The music glove refers to the mitts championed by Imogen Heap.

future.

The next stop on our journey revealed that the printed media industry is at a particularly interesting point in time. The prominence of digital media has been rapidly increasing, while the popularity of more traditional paper-based products has become less pronounced. As such, many companies in this industry are frantically working on transitioning their products to the digital world and adapting their business models to the 21st century.

By the end of our discussion, it became clear that innovation revolving around e-books, e-magazines, and the streamlining of distribution models are a few of the most promising areas for growth in this industry over the next decade.

Finally, we looked at a few other developments in entertainment that didn't quite fit neatly into any of the other categories. For example, we considered how technological innovation may change the world of sports. One of the more intriguing developments discussed here was the idea of placing digital sensors in athletic equipment and on the field to allow for increased accuracy in officiating.

The use of tablet computers by coaches, players, and staff during the game was another hot topic discussed at length. This issue will likely continue to be a point of much contention for years to come.

Lastly, we explored the ongoing debate surrounding the role technology will be permitted to play in major professional sports leagues. Ultimately, the decisions made in this particular area could have the most profound impact on how technology impacts sports in the years ahead.

Of course, the topics discussed throughout this book are only a sample of the seemingly endless list of technological developments occurring around the world every day. Nevertheless, it's my sincere hope that at least one of the topics discussed in this text has sparked enough interest in you, dear reader, to inspire further research and investigation into the topic than what could be covered within these pages.

As was mentioned numerous times throughout, the pace of technological development moves quickly, and I encourage all readers to perform their own research on the topics identified in this book. You may be surprised by how much progress is made between when this book was written and when you read it!

Readers are also highly encouraged to visit the companion website of this book (*www.impactofinnovation.com*) for the latest and most exciting updates on the areas discussed.

Relevant content submitted by readers will also be posted. So, if you happen to find an interesting article related to anything discussed in this book, please send it over using the email address found on the website – I'd love to see it!

The technological progress being made every day around the world is absolutely marvelous. We are truly living in an incredible time. When I started writing this book, I never thought I would find so many interesting developments in so many areas of the entertainment industry. Nevertheless, after researching countless topics related to technological advancement in the entertainment industry, I'm confident that one of the only things more surprising than our current state of technological development is the outlook of where our technology may lead us in the years ahead. As this book has shown, some innovations will clearly result in a better tomorrow, while others come with serious questions that will need addressed before they are introduced to society. Fortunately, there's plenty of time left to answer these questions and guide these technologies towards a better future for everyone.

There is no denying that the world of entertainment is rapidly changing, but where it's headed is far less of a certainty. Ultimately, the answer to that question is up to us.